Algebra For Grade 8

First Edition

Adam J. Wick

Independent Publishing Book

Preface

Algebra forms a crucial part of mathematics competitions, and its importance grows as students progress through their academic journey. For students in Grade 8, mastering algebra not only lays a solid foundation for future mathematical challenges but also opens up a world of exciting problem-solving opportunities. This book, **Mathematics Competitions Preparation, Volume 1: Algebra for Grade 8**, is tailored specifically for middle school students who are eager to excel in math competitions and build a strong understanding of algebraic concepts.

The goal of this book is to introduce Grade 8 students to the core ideas of algebra that are commonly encountered in math competitions at their level. It covers a variety of topics such as algebraic manipulation, equations, inequalities, and polynomials, focusing on methods that are both accessible and effective for students in this age group. Through a combination of clear explanations, illustrative examples, and challenging practice problems, this book aims to prepare students for the types of algebra problems they will face in contests.

While designed for competition preparation, this book can be used by students of varying skill levels. Beginners will find the content approachable, with step-by-step guidance to help them grasp each topic, while more advanced students will benefit from the challenging problems that encourage them to think creatively and critically. Each chapter is structured to build confidence and foster problem-solving skills, ensuring students are well-prepared for competition-level questions.

The practice problems provided at the end of each chapter are varied in difficulty, ranging from basic drills to more advanced problems that require deeper thought and strategic approaches. This range allows students to develop their skills gradually while pushing the boundaries of their mathematical thinking. Additionally, solutions are provided to help students learn from their mistakes and reinforce their understanding.

Beyond preparing students for competitions, this book aims to ignite a passion for mathematics. Algebra is not just about solving equations—it is about exploring patterns, discovering relationships, and understanding how different parts of mathematics fit together. Through this book, I hope to inspire curiosity, encourage persistence, and help students find the joy in solving even the most challenging problems.

As you work through this book, remember that success in mathematics competitions requires both practice and perseverance. Each problem you solve brings you one step closer to mastering algebra, and each challenge you face is an opportunity to grow as a mathematician. Keep an open mind, enjoy the process of learning, and never hesitate to explore new strategies and ideas.

I hope this book serves as a valuable resource in your preparation for mathematics competitions, and that it supports you in achieving your goals. Whether you are new to algebra or looking to refine your skills, I wish you the best of luck in your mathematical journey.

Adam J. Wick

Table of Contents

1 Operations on Numbers   **1**
1.1 Basic Laws of Addition and Multiplication   1
    1.1.1 Commutative Law   1
    1.1.2 Associative Law   1
    1.1.3 Distributive Law   2
1.2 Sums That Appeared In Most Competitions   4

2 Polynomials   **31**
2.1 Monomials   31
    2.1.1 Definition   31
    2.1.2 Degree   32
    2.1.3 Like Terms   32
    2.1.4 Addition and Subtraction of Like Terms . . .   33
    2.1.5 Multiplication of Monomials   34
2.2 Polynomials   35
    2.2.1 Definition   35
    2.2.2 Degree   35
    2.2.3 Operations on Polynomials   36

3 Methods in Factorization   **47**
3.1 Common Factor Methods   47
3.2 Splitting Middle Terms   49

4 Basic Algebraic Identities For Factorization   **99**
4.1 Perfect Square or Perfect Cube Identities   99
    4.1.1 Factoring Identities   105

5 Solutions   **113**

6 Square Roots — 119
6.1 Absolute Value 119
6.1.1 Definition 119
6.1.2 Properties of Absolute Values 120
6.2 Square Roots 120
6.3 Properties of Square Roots 122
6.4 Comparing Radical 124
6.5 Rationalize the Denominator 125
6.6 Fraction in Form $\frac{?}{\sqrt[n]{a}}$ 125
6.6.1 Fraction in the Form of $\frac{?}{\sqrt{a}+\sqrt{b}}$ or $\frac{?}{\sqrt{a}-\sqrt{b}}$ . . 127
6.7 Simplify Expressions in the Form of $\sqrt{a+2\sqrt{b}}$ . . . 130

7 Linear Equations in One Variable — 171
7.1 Defintions 171
7.2 Operation Properties of Equality 172
7.3 How to Solve It 173

8 Linear Inequalities With One Variable — 187
8.1 Defintion 187
8.2 Properties of Inequality 187
8.3 How To Solve It 188

9 Quadratic Equations — 193
9.1 Definition 193
9.2 How To Solve It 194
9.2.1 Product Equals Zero 194
9.2.2 Quadratic In Form: $x^2 = a$, where a is a positive real number 196
9.2.3 Completing Square 197
9.2.4 Solve Quadratic Equations By Discriminant . 199

Chapter 1

Operations on Numbers

We begin the first chapter of this book by introducing the basic laws of addition and multiplication. We believe that readers used to see all of those laws. However, we try to generalize all of them for all real numbers. We just want to make sure that readers understand the general concept of both operations.

1.1 Basic Laws of Addition and Multiplication

1.1.1 Commutative Law

For all real numbers a and b, we obtain
$$a + b = b + a$$
$$\text{and} \quad ab = ba.$$

1.1.2 Associative Law

For all real numbers a, b and c, we obtain
$$(a + b) + c = a + (b + c) = a + b + c$$
$$\text{and} \quad (ab)c = a(bc) = abc.$$

1.1.3 Distributive Law

For all real numbers a, b and c, we obtain

$$a(b+c) = (b+c)a = ab + ac.$$

Notice that subtraction and division are derived from addition and multiplication respectively.
That is, $a - b = a + (-b)$ and $\dfrac{a}{b} = a \times \dfrac{1}{b}$.

> **Example 1**
> Show that $(a+b)(c+d) = ac + ad + bc + bd$ for all real numbers a, b, c and d.

Solution. Show that $(a+b)(c+d) = ac + ad + bc + bd$.
Using distributive law, it implies that

$$\begin{aligned}(a+b)(c+d) &= (a+b)\,c + (a+b)\,d \\ &= ac + bc + ad + bd.\end{aligned}$$

> **Example 2**
> For all real numbers a and b, prove that
> $$(a+b)^2 = a^2 + 2ab + b^2.$$

Solution. Prove that $(a+b)^2 = a^2 + 2ab + b^2$.
We have

$$\begin{aligned}(a+b)^2 &= (a+b)(a+b) \\ &= a^2 + ab + ab + b^2 \\ &= a^2 + 2ab + b^2.\end{aligned}$$

Therefore, $(a+b)^2 = a^2 + 2ab + b^2$.

Remark 1. Taking $a = x$ and $b = 1$, we obtain

$$(x+1)^2 = x^2 + 2x + 1.$$

1.1. Basic Laws of Addition and Multiplication

Example 3

For all real numbers a and b, prove that
$$(a+b)^3 = a^3 + 3a^2b + 3ab^2 + b^3.$$

Solution. Prove that $(a+b)^3 = a^3 + 3a^2b + 3ab^2 + b^3$.
From the previous example, we have $(a+b)^2 = a^2 + 2ab + b^2$.
It follows that
$$\begin{aligned}(a+b)^3 &= (a+b)^2(a+b) \\ &= \left(a^2 + 2ab + b^2\right)(a+b) \\ &= a^3 + a^2b + 2a^2b + 2ab^2 + ab^2 + b^3 \\ &= a^3 + 3a^2b + 3ab^2 + b^3.\end{aligned}$$

Therefore, $(a+b)^3 = a^3 + 3a^2b + 3ab^2 + b^3$.

Remark 2. Taking $a = x$ and $b = 1$, we obtain
$$(x+1)^3 = x^3 + 3x^2 + 3x + 1.$$

Example 4

For all real numbers a and b, prove that
$$(a+b)^4 = a^4 + 4a^3b + 6a^2b^2 + 4ab^3 + b^4.$$

Solution. Prove that $(a+b)^4 = a^4 + 4a^3b + 6a^2b^2 + 4ab^3 + b^4$.
From the previous example, we have
$$(a+b)^3 = a^3 + 3a^2b + 3ab^2 + b^3.$$

It follows that
$$\begin{aligned}(a+b)^4 &= (a+b)^3(a+b) \\ &= \left(a^3 + 3a^2b + 3ab^2 + b^3\right)(a+b) \\ &= a^4 + a^3b + 3a^3b + 3a^2b^2 + 3a^2b^2 + 3ab^3 + ab^3 + b^4 \\ &= a^4 + 4a^3b + 6a^2b^2 + 4ab^3 + b^4.\end{aligned}$$

Therefore, $(a+b)^4 = a^4 + 4a^3b + 6a^2b^2 + 4ab^3 + b^4$.

Remark 3. Taking $a = x$ and $b = 1$, we obtain
$$(x+1)^4 = x^4 + 4x^3 + 6x^2 + 4x + 1.$$

1.2 Sums That Appeared In Most Competitions

The identities that are used in the following proofs:

1. $(k+1)^3 = k^3 + 3k^2 + 3k + 1$;

2. $(k+1)^4 = k^4 + 4k^3 + 6k^2 + 4k + 1$.

> **Example 5**
> Compute $S = 1 + 2 + 3 + \ldots + n$.

Solution. Compute S.
We have $\begin{cases} S = 1 + 2 + 3 + \ldots + (n-1) + n \\ S = n + (n-1) + \ldots + 2 + 1 \end{cases}$.
Adding the equalities, we obtain

$$2S = \underbrace{(n+1) + (n+1) + \ldots + (n+1)}_{n \quad \text{terms}} = n(n+1).$$

Consequently, $S = \dfrac{n(n+1)}{2}$.

> **Example 6**
> Use $1 + 2 + 3 + \ldots + n = \dfrac{n(n+1)}{2}$, find the values of the following expressions:
>
> 1. $S_1 = 10 + 11 + 12 + \ldots + 100$;
> 2. $S_2 = 15 + 16 + 17 + \ldots + 105$;
> 3. $S_3 = 20 + 21 + 22 + \ldots + 110$;
> 4. $S_4 = 2 + 4 + 6 + 8 + \ldots + 200$;
> 5. $S_5 = 1 + 3 + 5 + 7 + \ldots + 99$.

Solution. Find the values of the following expressions:

1.2. Sums That Appeared In Most Competitions

1. $S_1 = 10 + 11 + 12 + ... + 100$
   We have
   $$\begin{aligned} S_1 &= 10 + 11 + 12 + ... + 100 \\ &= (1 + 2 + 3 + ... + 100) - (1 + 2 + 3 + ... + 9) \\ &= \frac{100(100+1)}{2} - \frac{9(9+1)}{2} \\ &= 50(101) - 5 \times 9 \\ &= 5050 - 45 \\ &= 5005. \end{aligned}$$
   Therefore, $S_1 = 5005$.

2. $S_2 = 15 + 16 + 17 + ... + 105$
   We have
   $$\begin{aligned} S_2 &= 15 + 16 + 17 + ... + 105 \\ &= (1 + 2 + 3 + ... + 105) - (1 + 2 + 3 + ... + 14) \\ &= \frac{105(105+1)}{2} - \frac{14(14+1)}{2} \\ &= 53 \times 105 - 7 \times 15 \\ &= 5565 - 105 \\ &= 5460. \end{aligned}$$
   Therefore, $S_2 = 5460$.

3. $S_3 = 20 + 21 + 22 + ... + 110$
   We have
   $$\begin{aligned} S_3 &= 20 + 21 + 22 + ... + 110 \\ &= (1 + 2 + 3 + ... + 110) - (1 + 2 + 3 + ... + 19) \\ &= \frac{110(110+1)}{2} - \frac{19(19+1)}{2} \\ &= 55(111) - 10 \times 19 \\ &= 6105 - 190 \\ &= 5915. \end{aligned}$$
   Therefore, $S_3 = 5915$.

4. $S_4 = 2 + 4 + 6 + 8 + \ldots + 200$
   We have
   $$\begin{aligned} S_4 &= 2 + 4 + 6 + 8 + \ldots + 200 \\ &= 2(1 + 2 + 3 + 4 + \ldots + 100) \\ &= 2 \times \frac{100(100+1)}{2} \\ &= 100 \times 101 \\ &= 10100. \end{aligned}$$

   Therefore, $S_2 = 10100$.

5. $S_5 = 1 + 3 + 5 + 7 + \ldots + 99$.
   We have
   $$\begin{aligned} S_5 &= 1 + 3 + 5 + 7 + \ldots + 99 \\ &= (2 + 4 + 6 + 8 + \ldots + 100) - \Big(\underbrace{1 + 1 + \ldots + 1}_{50}\Big) \\ &= 2(1 + 2 + 3 + \ldots + 50) - 50 \\ &= 2 \times \frac{50(50+1)}{2} - 50 \\ &= 50 \times 51 + 50 \\ &= 2550 - 50 \\ &= 2500. \end{aligned}$$

   Therefore, $S_5 = 2500$.

> **Example 7**
>
> Compute $S = 1^2 + 2^2 + 3^2 + \ldots + n^2$.

Solution. For all integers k, we have
$$(k+1)^3 = k^3 + 3k^2 + 3k + 1$$
or
$$(k+1)^3 - k^3 = 3k^2 + 3k + 1.$$
It implies that
$$2^3 - 1^3 = 3(1^2) + 3(1) + 1;$$

1.2. Sums That Appeared In Most Competitions

$$3^3 - 2^3 = 3(2^2) + 3(2) + 1;$$
$$4^3 - 3^3 = 3(3^2) + 3(3) + 1;$$
$$\vdots$$
$$\text{and} \quad (n+1)^3 - n^3 = 3(n^2) + 3(n) + 1.$$

Adding all of the above identities, we obtain

$$(n+1)^3 - 1 = 3\left(1^2 + 2^2 + 3^2 + \ldots + n^2\right) + 3(1+2+3+\ldots+n) + n$$

$$(n+1)^3 - 1 = 3S + \frac{3n(n+1)}{2} + n$$

$$3S = (n+1)^3 - \frac{3n(n+1)}{2} - (n+1)$$

$$= (n+1)\left[(n+1)^2 - \frac{3n}{2} - 1\right]$$

$$= (n+1)\left[\frac{2(n+1)^2 - 3n - 2}{2}\right]$$

$$= (n+1)\left[\frac{2(n^2 + 2n + 1) - 3n - 2}{2}\right]$$

$$= (n+1)\left(\frac{2n^2 + 4n + 2 - 3n - 2}{2}\right)$$

$$= \frac{(n+1)(2n^2 + 4n + 2 - 3n - 2)}{2}$$

$$= \frac{(n+1)(2n^2 + n)}{2}$$

$$= \frac{n(n+1)(2n+1)}{2}.$$

Thus, $S = \dfrac{n(n+1)(2n+1)}{6}$.

Example 8

Using $1^2 + 2^2 + 3^2 + \ldots + n^2 = \dfrac{n(n+1)(2n+1)}{6}$, evaluate the following expressions:

1. $S_1 = 6^2 + 7^2 + 8^2 + \ldots + 20^2;$

2. $S_2 = 11^2 + 12^2 + 13^2 + ... + 20^2$;
3. $S_3 = 2^2 + 4^2 + 6^2 + ... + 20^2$;
4. $S_4 = 3^2 + 6^2 + 9^2 + ... + 30^2$.

Solution. Evaluate the following expressions:

1. $S_1 = 6^2 + 7^2 + 8^2 + ... + 20^2$
   We have
   $$\begin{aligned} S_1 &= 6^2 + 7^2 + 8^2 + ... + 20^2 \\ &= \left(1^2 + 2^2 + 3^2 + ... + 20^2\right) - \left(1^2 + 2^2 + 3^2 + 4^2 + 5^2\right) \\ &= \frac{20(20+1)(2 \times 20 + 1)}{6} - \frac{5(5+1)(2 \times 5 + 1)}{6} \\ &= \frac{20 \times 21 \times 41}{6} - \frac{5 \times 6 \times 11}{6} \\ &= 2870 - 55 \\ &= 2815. \end{aligned}$$

   Therefore, $S_1 = 2815$.

2. $S_2 = 11^2 + 12^2 + 13^2 + ... + 20^2$
   We have
   $$\begin{aligned} S_2 &= 11^2 + 12^2 + 13^2 + ... + 20^2 \\ &= \left(1^2 + 2^2 + 3^2 + ... + 20^2\right) - \left(1^2 + 2^2 + 3^2 + ... + 10^2\right) \\ &= \frac{20(20+1)(2 \times 20 + 1)}{6} - \frac{10(10+1)(2 \times 10 + 1)}{6} \\ &= \frac{20 \times 21 \times 41}{6} - \frac{10 \times 11 \times 21}{6} \\ &= 2870 - 385 \\ &= 2485. \end{aligned}$$

   Therefore, $S_2 = 2485$.

3. $S_3 = 2^2 + 4^2 + 6^2 + ... + 20^2$
   We have
   $$S_3 = 2^2 + 4^2 + 6^2 + ... + 20^2$$

1.2. Sums That Appeared In Most Competitions

$$= 2^2 \left(1^2 + 2^2 + 3^2 + ... + 10^2\right)$$
$$= 4 \times \frac{10(10+1)(2 \times 10 + 1)}{6}$$
$$= 4 \times \frac{10 \times 11 \times 21}{6}$$
$$= 4 \times 385$$
$$= 1540.$$

Therefore, $S_3 = 1540$.

4. $S_4 = 3^2 + 6^2 + 9^2 + ... + 30^2$
We have

$$S_3 = 3^2 + 6^2 + 9^2 + ... + 30^2$$
$$= 3^2 \left(1^2 + 2^2 + 3^2 + ... + 10^2\right)$$
$$= 9 \times \frac{10(10+1)(2 \times 10 + 1)}{6}$$
$$= 9 \times \frac{10 \times 11 \times 21}{6}$$
$$= 3465$$
$$= 1540.$$

Therefore, $S_4 = 1540$.

> **Example 9**
> Compute $S = 1^3 + 2^3 + 3^3 + ... + n^3$.

Solution. Compute S.
For all positive integers k, we have

$$(k+1)^4 = k^4 + 4k^3 + 6k^2 + 4k + 1$$

or

$$(k+1)^4 - k^4 = 4k^3 + 6k^2 + 4k + 1.$$

It follows that

$$2^4 - 1^4 = 4\left(1^3\right) + 6\left(1^2\right) + 4(1) + 1;$$
$$3^4 - 2^4 = 4\left(2^3\right) + 6\left(2^2\right) + 4(2) + 1;$$

$$4^4 - 3^4 = 4(3^3) + 6(3^2) + 4(3) + 1;$$

$$\vdots$$

and $(n+1)^4 - n^4 = 4(n^3) + 6(n^2) + 4(n) + 1.$

Adding all of the above identities, we obtain

$$(n+1)^4 - 1 = 4\left(1^3 + 2^3 + 3^3 + \ldots + n^3\right) + 6\left(1^2 + 2^2 + 3^2 + \ldots + n^2\right)$$
$$+ 4(1 + 2 + 3 + \ldots + n) + n$$

$$(n+1)^4 - 1 = 4S + \frac{6n(n+1)(2n+1)}{6} + \frac{4n(n+1)}{2} + n$$

$$(n+1)^4 - 1 = 4S + n(n+1)(2n+1) + 2n(n+1) + n$$

$$4S = (n+1)^4 - n(n+1)(2n+1) - 2n(n+1) - (n+1)$$
$$= (n+1)\left[(n+1)^3 - n(2n+1) - 2n - 1\right]$$
$$= (n+1)\left(n^3 + 3n^2 + 3n + 1 - 2n^2 - n - 2n - 1\right)$$
$$= (n+1)\left(n^3 + n^2\right)$$
$$= n^2(n+1)(n+1)$$
$$= [n(n+1)]^2.$$

Therefore, $S = \left[\dfrac{n(n+1)}{2}\right]^2.$

Example 10

Compute $S = \dfrac{1}{1 \times 2} + \dfrac{1}{2 \times 3} + \dfrac{1}{3 \times 4} + \ldots + \dfrac{1}{n(n+1)}.$

Solution. Compute S.
Observe that

$$\frac{1}{k(k+1)} = \frac{k+1-k}{k(k+1)}$$
$$= \frac{k+1}{k(k+1)} - \frac{k}{k(k+1)}$$
$$= \frac{1}{k} - \frac{1}{k+1}.$$

1.2. Sums That Appeared In Most Competitions

It implies that
$$\frac{1}{1 \times 2} = 1 - \frac{1}{2};$$
$$\frac{1}{2 \times 3} = \frac{1}{2} - \frac{1}{3};$$
$$\frac{1}{3 \times 4} = \frac{1}{3} - \frac{1}{4};$$
$$\vdots$$
and
$$\frac{1}{n(n+1)} = \frac{1}{n} - \frac{1}{n+1}.$$

Adding all of the above identities, we obtain
$$S = 1 - \frac{1}{n+1} = \frac{n+1-1}{n+1} = \frac{n}{n+1}.$$

Example 11

Evaluate the following sums:

1. $S_1 = 1 + \dfrac{1}{1+2} + \dfrac{1}{1+2+3} + \ldots + \dfrac{1}{1+2+\ldots+n};$

2. $S_2 = \dfrac{\dfrac{1}{(1 \times 3)^2} + \dfrac{2}{(3 \times 5)^2} + \dfrac{3}{(5 \times 7)^2} + \ldots + \dfrac{n}{[(2n-1)(2n+1)]^2}};$

3. $S_3 = \dfrac{1^2}{1 \times 3} + \dfrac{2^2}{3 \times 5} + \dfrac{3^2}{5 \times 7} + \ldots + \dfrac{n^2}{(2n-1)(2n+1)}.$

Solution. Compute:

1. $S_1 = 1 + \dfrac{1}{1+2} + \dfrac{1}{1+2+3} + \ldots + \dfrac{1}{1+2+\ldots+n}$
 Using $1 + 2 + 3 + \ldots + n = \dfrac{n(n+1)}{2}$, it follows that
$$\frac{1}{1+2+3+\ldots+n} = \frac{1}{\dfrac{n(n+1)}{2}} = \frac{2}{n(n+1)}.$$

We have

$$S = 1 + \frac{2}{2(2+1)} + \frac{2}{3(3+1)} + \frac{2}{4(4+1)} + \ldots + \frac{2}{n(n+1)}$$

$$= 1 + 2\left[\frac{1}{2\times 3} + \frac{1}{3\times 4} + \frac{1}{4\times 5} + \ldots + \frac{1}{n(n+1)}\right]$$

$$= 1 + 2\left[\left(\frac{1}{2} - \frac{1}{3}\right) + \left(\frac{1}{3} - \frac{1}{4}\right) + \left(\frac{1}{4} - \frac{1}{5}\right) + \ldots + \left(\frac{1}{n} - \frac{1}{n+1}\right)\right]$$

$$= 1 + 2\left(\frac{1}{2} - \frac{1}{n+1}\right)$$

$$= 1 + 1 - \frac{2}{n+1}$$

$$= 2 - \frac{2}{n+1}$$

$$= \frac{2n + 2 - 2}{n+1}$$

$$= \frac{2n}{n+1}.$$

2. $S_2 = \dfrac{1}{(1\times 3)^2} + \dfrac{2}{(3\times 5)^2} + \dfrac{3}{(5\times 7)^2} + \ldots + \dfrac{n}{[(2n-1)(2n+1)]^2}$

Observe that $\dfrac{k}{[(2k-1)(2k+1)]^2} = \dfrac{1}{8}\left[\dfrac{1}{(2k-1)^2} - \dfrac{1}{(2k+1)^2}\right].$

Consequently,

$$\frac{1}{(1\times 3)^2} = \frac{1}{8}\left(\frac{1}{1^2} - \frac{1}{3^2}\right);$$

$$\frac{1}{(3\times 5)^2} = \frac{1}{8}\left(\frac{1}{3^2} - \frac{1}{5^2}\right);$$

$$\frac{1}{(5\times 7)^2} = \frac{1}{8}\left(\frac{1}{5^2} - \frac{1}{7^2}\right);$$

$$\vdots$$

and $\dfrac{n}{[(2n-1)(2n+1)]^2} = \dfrac{1}{8}\left[\dfrac{1}{(2n-1)^2} - \dfrac{1}{(2n+1)^2}\right].$

1.2. Sums That Appeared In Most Competitions

Adding all of the above identities, we obtain

$$S = \frac{1}{8}\left[1 - \frac{1}{(2n+1)^2}\right]$$

$$= \frac{(2n+1)^2 - 1}{8(2n+1)^2}$$

$$= \frac{4n^2 + 4n + 1 - 1}{8(2n+1)^2}$$

$$= \frac{4n(n+1)}{8(2n+1)^2} = \frac{n(n+1)}{2(2n+1)^2}.$$

3. $S_3 = \dfrac{1^2}{1 \times 3} + \dfrac{2^2}{3 \times 5} + \dfrac{3^2}{5 \times 7} + \ldots + \dfrac{n^2}{(2n-1)(2n+1)}$

Observe that $\dfrac{k^2}{(2k-1)(2k+1)} = \dfrac{1}{4}\left(\dfrac{k}{2k-1} + \dfrac{k}{2k+1}\right).$

We obtain

$$\frac{1^2}{1 \times 3} = \frac{1}{4}\left(\frac{1}{1} + \frac{1}{3}\right);$$

$$\frac{2^2}{3 \times 5} = \frac{1}{4}\left(\frac{2}{3} + \frac{2}{5}\right);$$

$$\frac{3^2}{5 \times 7} = \frac{1}{4}\left(\frac{3}{5} + \frac{3}{7}\right);$$

$$\vdots$$

and $\quad \dfrac{n^2}{(2n-1)(2n+1)} = \dfrac{1}{4}\left(\dfrac{n}{2n-1} + \dfrac{n}{2n+1}\right).$

Adding all of the above identities, we obtain

$$S_3 = \frac{1}{4}\left(\underbrace{1 + 1 + \ldots + 1}_{n \text{ terms}} + \frac{n}{2n+1}\right)$$

$$= \frac{1}{4}\left(n + \frac{n}{2n+1}\right)$$

$$= \frac{1}{4}\left(\frac{2n^2 + n + n}{2n+1}\right)$$

$$= \frac{1}{4}\left(\frac{2n^2 + 2n}{2n+1}\right)$$

$$= \frac{1}{4}\left[\frac{2n(n+1)}{2n+1}\right]$$
$$= \frac{n(n+1)}{2(2n+1)}.$$

Exercises

Problem 1. Using the law of addition and multiplication, evaluate the following expressions:

1. $a + a + a$;
2. $3a + a + a$;
3. $4a - a - a$;
4. $7a - a - a$;
5. $8a - 2a - 4a$;
6. $(3a + b) + (3b + a)$;
7. $(a - b) + (4a - 4b)$;
8. $(3a - b) + (4a - 2b)$.

Problem 2. Expand the following expressions:

1. $(a + 1)(b + 1)$
2. $(2a - 1)(2b - 1)$
3. $(3a - 1)(3b - 1)$
4. $a(a + 1) + b(b + 1)$
5. $(a + 1)b + (b + 1)a$
6. $a(2a - b) + b(2b - a)$
7. $a(b + c) + b(c + a) + c(a + b)$
8. $a(b - c) + b(c - a) + c(a - b)$

Chapter 1. Operations on Numbers

Problem 3. What is the value of the following product?

$$P = \left(1 - \frac{1}{2}\right)\left(1 - \frac{1}{3}\right)\left(1 - \frac{1}{4}\right)\left(1 - \frac{1}{5}\right)\left(1 - \frac{1}{6}\right)\left(1 - \frac{1}{7}\right).$$

Problem 4. What is the value of the following expression?

$$P = \left(1 - \frac{1}{2}\right)\left(1 - \frac{1}{3}\right)\left(1 - \frac{1}{4}\right)\cdots\left(1 - \frac{1}{n}\right).$$

Problem 5. What is the value of the following product?

$$P = \left(1 + \frac{2}{3}\right)\left(1 + \frac{2}{5}\right)\left(1 + \frac{2}{7}\right)\left(1 + \frac{1}{2}\right)\left(1 + \frac{1}{3}\right)\left(1 + \frac{1}{4}\right).$$

Problem 6. What is the value of the following product?

$$P = \left(1 + \frac{1}{2}\right)\left(1 + \frac{1}{3}\right)\left(1 + \frac{1}{4}\right)\cdots\left(1 + \frac{1}{n}\right).$$

Problem 7. Evaluate

$$S = 1^2 - 2^2 + 3^2 - 4^2 + \ldots + 2023^2 - 2024^2.$$

Problem 8. 1. For all positive integers k, prove that

$$\frac{2k+1}{[k(k+1)]^2} = \frac{1}{k^2} - \frac{1}{(k+1)^2}.$$

2. For all positive integers n, evaluate

$$S = \frac{3}{(1 \times 2)^2} + \frac{5}{(2 \times 3)^2} + \frac{7}{(3 \times 4)^2} + \ldots + \frac{2n+1}{[n(n+1)]^2}.$$

Problem 9. 1. For all positive integers k, prove that

$$\frac{6(1 + 2 + 3 + \ldots + k) + 1}{[k(k+1)]^3} = \frac{1}{k^3} - \frac{1}{(k+1)^3}.$$

2. For all positive integers n, evaluate

$$S = \frac{6(1) + 1}{(1 \times 2)^2} + \frac{6(1+2) + 1}{(2 \times 3)^2} + \ldots + \frac{6(1+2+\ldots+n)}{[n(n+1)]^2}.$$

1.2. Sums That Appeared In Most Competitions

Problem 10. For all positive integers n, evaluate
$$S = 1 + \frac{1}{1+2} + \frac{1}{1+2+3} + \ldots + \frac{1}{1+2+3+\ldots+n}.$$

Problem 11. For all positive integers $n \geq 2$, evaluate
$$S = \frac{2^2+1}{2^2-1} + \frac{3^2+1}{3^2-1} + \frac{4^2+1}{4^2-1} + \ldots + \frac{n^2+1}{n^2-1}.$$

Problem 12. For all positive integers n, evaluate
$$S = \frac{1}{1+1^2+1^4} + \frac{2}{1+2^2+2^4} + \ldots + \frac{n}{1+n^2+n^4}.$$

Solutions

Problem 1. Using the law of addition and multiplication, evaluate the following expressions:

1. $a + a + a$;
2. $3a + a + a$;
3. $4a - a - a$;
4. $7a - a - a$;
5. $8a - 2a - 4a$;
6. $(3a + b) + (3b + a)$;
7. $(a - b) + (4a - 4b)$;
8. $(3a - b) + (4a - 2b)$.

Solution. Using the law of addition and multiplication, evaluate the following expressions:

1. $a + a + a = (1 + 1 + 1)a = 3a$
2. $3a + a + a = (3 + 1 + 1)a = 5a$
3. $4a - a - a = (4 - 1 - 1)a = 2a$
4. $7a - a - a = (7 - 1 - 1)a = 5a$
5. $8a - 2a - 4a = (8 - 2 - 4)a = 2a$
6. $(3a + b) + (3b + a) = 3a + b + 3b + a = 4a + 4b$
7. $(a - b) + (4a - 4b) = a - b + 4a - 4b = 5a - 5b$

8. $(3a - b) + (4a - 2b) = 3a - b + 4a - 2b = 7a - 3b$

Problem 2. Expand the following expressions:

1. $(a + 1)(b + 1)$
2. $(2a - 1)(2b - 1)$
3. $(3a - 1)(3b - 1)$
4. $a(a + 1) + b(b + 1)$
5. $(a + 1)b + (b + 1)a$
6. $a(2a - b) + b(2b - a)$
7. $a(b + c) + b(c + a) + c(a + b)$
8. $a(b - c) + b(c - a) + c(a - b)$

Solution. Expand the following expressions:

1. $(a + 1)(b + 1) = ab + a + b + 1$
2. $(2a - 1)(2b - 1) = 4ab - 2a - 2b + 1$
3. $(3a - 1)(3b - 1) = 9ab - 3a - 3b + 1$
4. $a(a + 1) + b(b + 1) = a^2 + a + b^2 + b = a^2 + b^2 + a + b$
5. $(a + 1)b + (b + 1)a = ab + b + ab + a = 2ab + a + b$
6. $a(2a - b) + b(2b - a) = 2a^2 - ab + 2b^2 - ab = 2a^2 + 2b^2 - 2ab$
7.
$$a(b + c) + b(c + a) + c(a + b)$$
$$= ab + ac + bc + ab + ca + bc$$
$$= 2ab + 2bc + 2ca.$$

8.
$$a(b - c) + b(c - a) + c(a - b)$$
$$= ab - ac + bc - ab + ca - bc$$
$$= 0.$$

1.2. Sums That Appeared In Most Competitions

Problem 3. What is the value of the following product?
$$P = \left(1-\frac{1}{2}\right)\left(1-\frac{1}{3}\right)\left(1-\frac{1}{4}\right)\left(1-\frac{1}{5}\right)\left(1-\frac{1}{6}\right)\left(1-\frac{1}{7}\right).$$

Solution. Evaluate P.
We have
$$P = \left(1-\frac{1}{2}\right)\left(1-\frac{1}{3}\right)\left(1-\frac{1}{4}\right)\left(1-\frac{1}{5}\right)\left(1-\frac{1}{6}\right)\left(1-\frac{1}{7}\right)$$
$$= \frac{1}{2} \times \frac{2}{3} \times \frac{3}{4} \times \frac{4}{5} \times \frac{5}{6} \times \frac{6}{7}$$
$$= \frac{1}{7}.$$

Therefore, $P = \frac{1}{7}$.

Problem 4. What is the value of the following expression?
$$P = \left(1-\frac{1}{2}\right)\left(1-\frac{1}{3}\right)\left(1-\frac{1}{4}\right)\ldots\left(1-\frac{1}{n}\right).$$

Solution. Evaluate P.
We have
$$P = \left(1-\frac{1}{2}\right)\left(1-\frac{1}{3}\right)\left(1-\frac{1}{4}\right)\ldots\left(1-\frac{1}{n}\right)$$
$$= \frac{1}{2} \times \frac{2}{3} \times \frac{3}{4} \times \ldots \times \frac{n-1}{n}$$
$$= \frac{1}{n}.$$

Therefore, $P = \frac{1}{n}$.

Problem 5. What is the value of the following product?
$$P = \left(1+\frac{2}{3}\right)\left(1+\frac{2}{5}\right)\left(1+\frac{2}{7}\right)\left(1+\frac{1}{2}\right)\left(1+\frac{1}{3}\right)\left(1+\frac{1}{4}\right).$$

Solution. Evaluate P.
We have
$$P = \left(1+\frac{2}{3}\right)\left(1+\frac{2}{5}\right)\left(1+\frac{2}{7}\right)\left(1+\frac{1}{2}\right)\left(1+\frac{1}{3}\right)\left(1+\frac{1}{4}\right)$$

$$= \frac{5}{3} \times \frac{7}{5} \times \frac{9}{7} \times \frac{3}{2} \times \frac{4}{3} \times \frac{5}{4}$$
$$= 3 \times \frac{5}{2}$$
$$= \frac{15}{2}.$$

Therefore, $P = \frac{15}{2}$.

Problem 6. What is the value of the following product?
$$P = \left(1 + \frac{1}{2}\right)\left(1 + \frac{1}{3}\right)\left(1 + \frac{1}{4}\right) \ldots \left(1 + \frac{1}{n}\right).$$

Solution. Evaluate P.
We have
$$P = \left(1 + \frac{1}{2}\right)\left(1 + \frac{1}{3}\right)\left(1 + \frac{1}{4}\right) \ldots \left(1 + \frac{1}{n}\right)$$
$$= \frac{3}{2} \times \frac{4}{3} \times \frac{5}{4} \times \ldots \times \frac{n+1}{n}$$
$$= \frac{n+1}{2}.$$

Therefore, $P = \frac{n+1}{2}$.

Problem 7. Evaluate
$$S = 1^2 - 2^2 + 3^2 - 4^2 + \ldots + 2023^2 - 2024^2.$$

Solution. Evaluate S.
We have
$$S = 1^2 - 2^2 + 3^2 - 4^2 + \ldots + 2023^2 - 2024^2$$
$$= \left(1^2 - 2^2\right) + \left(3^2 - 4^2\right) + \ldots + \left(2023^2 - 2024^2\right)$$
$$= (1-2)(1+2) + (3-4)(3+4) + \ldots$$
$$= -1 - 2 - 3 - 4 - \ldots - 2023 - 2024$$
$$= -(1 + 2 + 3 + 4 + \ldots + 2024)$$
$$= -\frac{2024(2024+1)}{2}$$
$$= -1012 \times 2025$$
$$= -2049300.$$

Therefore, $S = -2049300$.

1.2. Sums That Appeared In Most Competitions

Problem 8.   1. For all positive integers k, prove that
$$\frac{2k+1}{[k(k+1)]^2} = \frac{1}{k^2} - \frac{1}{(k+1)^2}.$$

2. For all positive integers n, evaluate
$$S = \frac{3}{(1\times 2)^2} + \frac{5}{(2\times 3)^2} + \frac{7}{(3\times 4)^2} + \dots + \frac{2n+1}{[n(n+1)]^2}.$$

Solution.   1. For all positive integers k, prove that
$$\frac{2k+1}{[k(k+1)]^2} = \frac{1}{k^2} - \frac{1}{(k+1)^2}.$$

For all positive integers k, we have
$$\frac{1}{k^2} - \frac{1}{(k+1)^2} = \frac{(k+1)^2 - k^2}{k^2(k+1)^2}$$
$$= \frac{k^2 + 2k + 1 - k^2}{[k(k+1)]^2}$$
$$= \frac{2k+1}{[k(k+1)]^2}.$$

Therefore, $\dfrac{2k+1}{[k(k+1)]^2} = \dfrac{1}{k^2} - \dfrac{1}{(k+1)^2}.$

2. For all positive integers n, evaluate
$$S = \frac{3}{(1\times 2)^2} + \frac{5}{(2\times 3)^2} + \frac{7}{(3\times 4)^2} + \dots + \frac{2n+1}{[n(n+1)]^2}.$$

From the above proof, we have
$$\frac{2k+1}{[k(k+1)]^2} = \frac{1}{k^2} - \frac{1}{(k+1)^2}.$$

Taking $k = 1, 2, 3, \dots, n$, we obtain
$$\frac{3}{(1\times 2)^2} = \frac{1}{1^2} - \frac{1}{2^2};$$
$$\frac{5}{(2\times 3)^2} = \frac{1}{2^2} - \frac{1}{3^2};$$

$$\frac{7}{(3\times 4)^2} = \frac{1}{3^2} - \frac{1}{4^2};$$

$$\vdots$$

and $\quad \dfrac{2n+1}{[n(n+1)]^2} = \dfrac{1}{n^2} - \dfrac{1}{(n+1)^2}.$

Adding all of the above identities, we obtain

$$S = \frac{1}{1^2} - \frac{1}{(n+1)^2}$$
$$= 1 - \frac{1}{(n+1)^2}$$
$$= \frac{(n+1)^2 - 1}{(n+1)^2}$$
$$= \frac{n^2 + 2n + 1 - 1}{(n+1)^2}$$
$$= \frac{n^2 + 2n}{(n+1)^2}.$$

Therefore, $S = \dfrac{n^2+2n}{(n+1)^2}.$

Problem 9. 1. For all positive integers k, prove that

$$\frac{6(1+2+3+\ldots+k)+1}{[k(k+1)]^3} = \frac{1}{k^3} - \frac{1}{(k+1)^3}.$$

2. For all positive integers n, evaluate

$$S = \frac{6(1)+1}{(1\times 2)^2} + \frac{6(1+2)+1}{(2\times 3)^2} + \ldots + \frac{6(1+2+\ldots+n)}{[n(n+1)]^2}.$$

Solution. 1. For all positive integers k, prove that

$$\frac{6(1+2+3+\ldots+k)+1}{[k(k+1)]^3} = \frac{1}{k^3} - \frac{1}{(k+1)^3}.$$

For all positive integers k, we have

$$\frac{1}{k^3} - \frac{1}{(k+1)^3} = \frac{(k+1)^3 - k^3}{k^3(k+1)^3}$$

24

1.2. Sums That Appeared In Most Competitions

$$= \frac{k^3 + 3k^2 + 3k + 1 - k^3}{[k(k+1)]^3}$$

$$= \frac{3k^2 + 3k + 1}{[k(k+1)]^3}$$

$$= \frac{3k(k+1) + 1}{[k(k+1)]^3}$$

$$= \frac{6 \times \frac{k(k+1)}{2} + 1}{[k(k+1)]^3}$$

$$= \frac{6(1 + 2 + \ldots + k) + 1}{[k(k+1)]^3}.$$

Therefore, $\dfrac{6(1 + 2 + 3 + \ldots + k) + 1}{[k(k+1)]^3} = \dfrac{1}{k^3} - \dfrac{1}{(k+1)^3}.$

2. For all positive integers n, evaluate

$$S = \frac{6(1) + 1}{(1 \times 2)^2} + \frac{6(1+2) + 1}{(2 \times 3)^2} + \ldots + \frac{6(1+2+\ldots+n)}{[n(n+1)]^2}.$$

From the above proof, we have

$$\frac{6(1) + 1}{(1 \times 2)^3} = \frac{1}{1^3} - \frac{1}{2^3};$$

$$\frac{6(1+2) + 1}{(2 \times 3)^3} = \frac{1}{2^3} - \frac{1}{3^3};$$

$$\frac{6(1+2+3) + 1}{(3 \times 4)^3} = \frac{1}{3^3} - \frac{1}{4^3};$$

$$\vdots$$

and $\dfrac{6(1+2+\ldots+n) + 1}{[n(n+1)]^3} = \dfrac{1}{n^3} - \dfrac{1}{(n+1)^3}.$

Adding all of the above identities, we obtain

$$S = \frac{1}{1^3} - \frac{1}{(n+1)^3}$$

$$= 1 - \frac{1}{(n+1)^3}.$$

$$= \frac{(n+1)^3 - 1}{(n+1)^3}$$
$$= \frac{n^3 + 3n^2 + 3n + 1 - 1}{(n+1)^3}$$
$$= \frac{n^3 + 3n^2 + 3n}{(n+1)^3}.$$

Therefore, $S = \dfrac{n^3 + 3n^2 + 3n}{(n+1)^3}$.

Problem 10. For all positive integers n, evaluate
$$S = 1 + \frac{1}{1+2} + \frac{1}{1+2+3} + \ldots + \frac{1}{1+2+3+\ldots+n}.$$

Solution. Evaluate S.
For all positive integers k, we have
$$\frac{1}{1+2+3+\ldots+k} = \frac{1}{\frac{k(k+1)}{2}}$$
$$= \frac{2}{k(k+1)}$$
$$= 2\left(\frac{1}{k} - \frac{1}{k+1}\right).$$

Taking $k = 1, 2, 3, \ldots, n$, we obtain
$$1 = 2\left(\frac{1}{1} - \frac{1}{2}\right);$$
$$\frac{1}{1+2} = 2\left(\frac{1}{2} - \frac{1}{3}\right);$$
$$\frac{1}{1+2+3} = 2\left(\frac{1}{3} - \frac{1}{4}\right);$$
$$\vdots$$
$$\text{and} \quad \frac{1}{1+2+3+\ldots+n} = 2\left(\frac{1}{n} - \frac{1}{n+1}\right).$$

Adding all of the above identities, we obtain
$$S = 2\left(\frac{1}{1} - \frac{1}{n+1}\right)$$

1.2. Sums That Appeared In Most Competitions

$$= 2\left(1 - \frac{1}{n+1}\right)$$
$$= 2 \times \frac{n+1-1}{n+1}$$
$$= \frac{2n}{n+1}.$$

Therefore, $S = \dfrac{2n}{n+1}$.

Problem 11. For all positive integers $n \geq 2$, evaluate
$$S = \frac{2^2+1}{2^2-1} + \frac{3^2+1}{3^2-1} + \frac{4^2+1}{4^2-1} + \ldots + \frac{n^2+1}{n^2-1}.$$

Solution. Evaluate S.
For all positive integers $k \geq 2$, we have
$$\frac{k^2+1}{k^2-1} = \frac{k^2-1+2}{k^2-1}$$
$$= 1 + \frac{2}{k^2-1}$$
$$= 1 + \frac{2}{(k-1)(k+1)}$$
$$= 1 + \left(\frac{1}{k-1} - \frac{1}{k+1}\right).$$

Taking $k = 2, 3, 4, \ldots, n$, we obtain
$$\frac{2^2+1}{2^2-1} = 1 + \left(\frac{1}{1} - \frac{1}{2}\right);$$
$$\frac{3^2+1}{3^2-1} = 1 + \left(\frac{1}{2} - \frac{1}{3}\right);$$
$$\frac{4^2+1}{4^2-1} = 1 + \left(\frac{1}{3} - \frac{1}{4}\right);$$
$$\vdots$$
and $\quad \dfrac{n^2+1}{n^2-1} = 1 + \left(\dfrac{1}{n} - \dfrac{1}{n+1}\right).$

Adding all of the above identities, we obtain
$$S = n + \left(\frac{1}{1} - \frac{1}{n+1}\right)$$

$$= n+1 - \frac{1}{n+1}$$
$$= \frac{(n+1)^2 - 1}{n+1}$$
$$= \frac{n^2 + 2n + 1 - 1}{n+1}$$
$$= \frac{n^2 + 2n}{n+1}.$$

Therefore, $S = \dfrac{n^2+2n}{n+1}$.

Problem 12. For all positive integers n, evaluate
$$S = \frac{1}{1+1^2+1^4} + \frac{2}{1+2^2+2^4} + \dots + \frac{n}{1+n^2+n^4}.$$

Solution. Evaluate S.
For all positive integers k, we have
$$\frac{k}{1+k^2+k^4} = \frac{k}{k^4+2k^2+1-k^2}$$
$$= \frac{k}{(k^2+1)^2 - k^2}$$
$$= \frac{k}{(k^2+1-k)(k^2+1+k)}$$
$$= \frac{1}{2}\left(\frac{1}{k^2-k+1} - \frac{1}{k^2+k+1}\right)$$
$$= \frac{1}{2}\left[\frac{1}{k(k-1)+1} - \frac{1}{k(k+1)+1}\right].$$

Taking $k = 1, 2, 3, \dots, n$, we obtain
$$\frac{1}{1+1^2+1^4} = \frac{1}{2}\left(\frac{1}{0\times 1+1} - \frac{1}{1\times 2+1}\right);$$
$$\frac{2}{1+2^2+2^4} = \frac{1}{2}\left(\frac{1}{1\times 2+1} - \frac{1}{2\times 3+1}\right);$$
$$\frac{3}{1+3^2+3^4} = \frac{1}{2}\left(\frac{1}{2\times 3+1} - \frac{1}{3\times 4+1}\right);$$
$$\vdots$$

1.2. Sums That Appeared In Most Competitions

and $\dfrac{n}{1+n^2+n^4} = \dfrac{1}{2}\left[\dfrac{1}{(n-1)n+1} - \dfrac{1}{n(n+1)+1}\right].$

Adding all of the above identities, we obtain

$$\begin{aligned} S &= \dfrac{1}{2}\left[\dfrac{1}{0 \times 1 + 1} - \dfrac{1}{n(n+1)+1}\right] \\ &= \dfrac{1}{2}\left(1 - \dfrac{1}{n^2+n+1}\right) \\ &= \dfrac{1}{2} \times \dfrac{n^2+n+1-1}{n^2+n+1} \\ &= \dfrac{n^2+n}{2(n^2+n+1)}. \end{aligned}$$

Therefore, $S = \dfrac{n^2+n}{2(n^2+n+1)}.$

Chapter 2

Polynomials

In the previous chapter, we introduced a basic concept in operations. Another basic concept in mathematics is Polynomial. In Mathematics, we usually see different type of polynomials. It is very useful to understand well about the terminology of polynomials. We begin this chapter by introducing the really basic parts of polynomials. It is called monomial.

2.1 Monomials

2.1.1 Definition

> **Definition 1**
> A monomial is a product of a number and a nonnegative power of variables.

> **Example 12**
> $2x$, x^2, $4x^7$, $4xy$, and $9xyz^2t$ are monomials.
> However, $\sqrt{x}$, $\dfrac{1}{x}$, and x^{-4} are not monomials.

Remark 4. Suppose that a is a constant. We can rewrite a as $a = ax^0$. That is, a is a monomial.

In general, all constants are monomials. The constant and variable parts of monomials are called coefficients and variables respectively.

> **Example 13**
>
> $2x^3$ is a monomial with 2 is the coefficient and x is the variable.

2.1.2 Degree

Degree of a monomial is the sum of all exponents of each variable of the monomial.

> **Example 14**
>
> Fill in the following table:
>
> | Monomial | Constant | Variable | Degree |
> |---|---|---|---|
> | $3x^4$ | | | |
> | $-7xy$ | | | |
> | $\frac{1}{2}x^2yz$ | | | |
> | $4abc$ | | | |
> | $\sqrt{2}x^3y^4z^5$ | | | |

Solution. Fill in the following table:

Monomial	Constant	Variable	Degree
$3x^4$	3	x	4
$-7xy$	-7	x, y	2
$\frac{1}{2}x^2yz$	$\frac{1}{2}$	x, y, z	4
$4abc$	4	a, b, c	3
$\sqrt{2}x^3y^4z^5$	$\sqrt{2}$	x, y, z	12

2.1.3 Like Terms

Two terms are like if they the same variable parts.

2.1. Monomials

> **Example 15**
> 1. $7xyz$ and $10xyz$ are called like terms
> 2. $6x^4$ and $13x^4$ are called like terms.

> **Example 16**
> Find like terms in the following expressions.
> $$7x^2, 3x^7, 4abc, 5abc^2, 6x^7, \frac{1}{2}x^2, -13ab^2c, -\sqrt{2}abc.$$

Solution. Like terms are:

- $7x^2$ and $\frac{1}{2}x^2$;
- $3x^7$ and $6x^7$;
- $4abc$ and $-\sqrt{2}abc$.

2.1.4 Addition and Subtraction of Like Terms

To add or subtract two or more like terms, we add or subtract only the coefficients of each term and keep the variable parts the same.

> **Example 17**
> Compute the following expressions:
> 1. $2x + 7x$;
> 2. $x^2 - 3x^2$;
> 3. $6xyz + 7xyz$;
> 4. $-11ax + 18ax$;
> 5. $7xy + 8xy - 3xy$.

Solution. Compute the following expressions:

1. $2x + 7x = (2+7)x = 9x$.
2. $x^2 - 3x^2 = (1-3)x^2 = -2x^2$.

3. $6xyz + 7xyz = (6+7)xyz = 13xyz$.

4. $-11ax + 18ax = (-11+18)ax = 7ax$.

5. $7xy + 8xy - 3xy = (7+8-3)xy = 12xy$.

2.1.5 Multiplication of Monomials

Exponent Properties:
It is really important to know about exponent properties before we start multiplying monomials. Here are some exponent properties that we have to remember.

1. $x^m \times x^n = x^{m+n}$;

2. $\dfrac{x^n}{x^m} = x^{n-m}$, where $x \neq 0$;

3. $(x^n)^m = x^{nm}$;

4. $(x \times y)^n = x^n \times y^n$;

5. $\left(\dfrac{x}{y}\right)^n = \dfrac{x^n}{y^n}$, where $y \neq 0$.

> **Example 18**
>
> Compute the following expressions:
>
> 1. $x^4 \times x^2$;
>
> 2. $\left(x^2 y^2\right)^2$;
>
> 3. $\dfrac{x^7 \times x^2}{x^6}$;
>
> 4. $\dfrac{(xyz)^3}{xy^2 z^3}$.

Solution. Compute the following expressions:

1. $x^4 \times x^2 = x^{4+2} = x^6$.

2. $\left(x^2 y^2\right)^2 = (x^2)^2 (y^2)^2 = x^4 y^4$.

3. $\dfrac{x^7 \times x^2}{x^6} = x^{7+2-6} = x^3$.

4. $\dfrac{(xyz)^3}{xy^2 z^3} = \dfrac{x^3 y^3 z^3}{xy^2 z^3} = x^2 y$.

2.2. Polynomials

> **Example 19**
> Prove that $a^0 = 1$ for all $a \neq 0$.

Solution. Observe that
$$\frac{a}{a} = 1 \quad \text{for all } a \neq 0$$
$$a^{1-1} = 1$$
$$a^0 = 1.$$

2.2 Polynomials

2.2.1 Definition

A polynomial is the sum of two or more monomials. Each monomial is called a term.

> **Example 20**
> $3x^2 - 2x + 1$, $x^3 - 4x + 6$, $x^7 + 6x - 3$, $x + 1$ are polynomials.

2.2.2 Degree

Degree of a polynomial is the highest degree of its monomial.

> **Example 21**
> $4x^3 - 2x + 1$ is a polynomial and its degree is 3.

> **Example 22**
> Fill in the following table.
>
> | Polynomial | Variable | Degree |
> |---|---|---|
> | $3x^7 - 5x + 1$ | | |
> | $x^2yz + xyz + 1$ | | |
> | $4x^3y - 3xy + 4x^5$ | | |
> | $ab + bc - abc$ | | |
> | $x^7y + xy^6 - 2$ | | |

Solution. Fill in the following table.

Polynomial	Variable	Degree
$3x^7 - 5x + 1$	x	7
$x^2yz + xyz + 1$	x, y, z	4
$4x^3y - 3xy + 4x^5$	x, y	5
$ab + bc - abc$	a, b, c	3
$x^7y + xy^6 - 2$	x, y	8

2.2.3 Operations on Polynomials

Addition and Subtraction of Polynomials

To add or subtract polynomials, we add or subtract like terms and arrange them in ascending order or descending order.

> **Example 23**
>
> Given that $A = x^2 - 3x + 3$, $B = -2x^2 + 4x - 1$ and $C = 5x^2 - 1$. Compute
>
> 1. $A + B$;
> 2. $A - C$;
> 3. $A - B + C$.

Solution. Compute

1. $A + B$

    We have

    $$\begin{aligned} A + B &= \left(x^2 - 3x + 3\right) + \left(-2x^2 + 4x - 1\right) \\ &= x^2 - 3x + 3 - 2x^2 + 4x - 1 \\ &= x^2 - 2x^2 - 3x + 4x + 3 - 1 \\ &= -x^2 + x + 2. \end{aligned}$$

2. $A - C$

    We have

    $$\begin{aligned} A - C &= \left(x^2 - 3x + 3\right) - \left(5x^2 - 1\right) \\ &= x^2 - 3x + 3 - 5x^2 + 1 \\ &= x^2 - 5x^2 - 3x + 3 + 1 \\ &= -4x^2 - 3x + 4. \end{aligned}$$

2.2. Polynomials

3. $A - B + C$

   We have

   $$\begin{aligned} A - B + C &= \left(x^2 - 3x + 3\right) - \left(-2x^2 + 4x - 1\right) + \left(5x^2 - 1\right) \\ &= x^2 - 3x + 3 + 2x^2 - 4x + 1 + 5x^2 - 1 \\ &= x^2 + 2x^2 + 5x^2 - 3x - 4x + 3 + 1 - 1 \\ &= 8x^2 - 7x + 3. \end{aligned}$$

Multiplication of Polynomials

To multiply two polynomials, we have to use distributive law that we have learnt in Chapter I. That is,

$$(a+b)(c+d) = ac + ad + bc + bd.$$

See the following examples.

Example 24

Expand the following expressions:

1. $(x+1)(x+7)$;
2. $(x-1)(x+3)$;
3. $(x-1)(x-6)$;
4. $(x+1)(x+2)(x+3)$;
5. $(x+1)(x+2)(x+3)(x+4)$.

Solution. Expand the following expressions:

1. $(x+1)(x+7)$

   We have

   $$\begin{aligned}(x+1)(x+7) &= x^2 + 7x + x + 7 \\ &= x^2 + 8x + 7.\end{aligned}$$

2. $(x-1)(x+3)$

   We have

   $$\begin{aligned}(x-1)(x+3) &= x^2 + 3x - x - 3 \\ &= x^2 + 2x - 3.\end{aligned}$$

3. $(x-1)(x-6)$
   We have
   $$(x-1)(x-6) = x^2 - 6x - x + 6$$
   $$= x^2 - 7x + 6.$$

4. $(x+1)(x+2)(x+3)$
   We have
   $$(x+1)(x+2)(x+3) = \left(x^2 + 2x + x + 2\right)(x+3)$$
   $$= \left(x^2 + 3x + 2\right)(x+3)$$
   $$= x^3 + 3x^2 + 3x^2 + 9x + 2x + 6$$
   $$= x^3 + 6x^2 + 11x + 6.$$

5. $(x+1)(x+2)(x+3)(x+4)$
   We have
   $$(x+1)(x+2)(x+3)(x+4)$$
   $$= [(x+1)(x+2)][(x+3)(x+4)]$$
   $$= \left(x^2 + 2x + x + 2\right)\left(x^2 + 4x + 3x + 12\right)$$
   $$= \left(x^2 + 3x + 2\right)\left(x^2 + 7x + 12\right)$$
   $$= x^4 + 7x^3 + 12x^2 + 3x^3 + 21x^2 + 36x + 2x^2 + 14x + 24$$
   $$= x^4 + 10x^3 + 35x^2 + 50x + 24.$$

Exercises

Problem 1.

Compute the following expressions:

1. $4x + 3x$;
2. $3x - 2x + 5x$;
3. $3x - 2(1-x) + 5x$;
4. $(3x + 4y) + (2x - y)$;
5. $3(x-y) + 4(y-x) + 5(2x-y)$;
6. $2(x+y) + 3(y-x) + 4y + 3x$;
7. $2(x+y+z) + 3(2x-y-z)$;
8. $4(x-y-z) + 3(z-y+x) - 2(2x-3y+4z)$.

Problem 2.
Given that $A = x^4 + x^3 + x^2 + x + 1$, $B = 2x^4 + 3x^3 + 5x + 6$ and $C = 3x^3 + 5x - 12$.
Compute

1. $A + B$;
2. $B + C$;
3. $C + A$;
4. $A + B + C$;
5. $A - B + C$;
6. $A + B - C$.

Problem 3. Expand the following expressions:

1. $(x+3)(x+4)$;

2. $(x+5)(x-6)$;

3. $(x+1)(x+2)(x-3)$;

4. $(x+y)(2x-y)$;

5. $x(x+1)(x+2)(x+8)$.

Problem 4.

Calculate the following expressions:

1. $(x+1)(x+2)+(x+2)(x+3)$;

2. $(x+1)(x^2+2)+(x^2+1)(x-1)$;

3. $(x+5)(x-1)+(x+1)(x-10)$;

4. $5(x+1)(x-3)+7x(3-x)$;

5. $(1-x)(2-x)(3-x)+(4-x)(5-x)(6-x)$.

Solutions

Problem 1.
Compute the following expressions:

1. $4x + 3x$;
2. $3x - 2x + 5x$;
3. $3x - 2(1-x) + 5x$;
4. $(3x + 4y) + (2x - y)$;
5. $3(x-y) + 4(y-x) + 5(2x-y)$;
6. $2(x+y) + 3(y-x) + 4y + 3x$;
7. $2(x+y+z) + 3(2x-y-z)$;
8. $4(x-y-z) + 3(z-y+x) - 2(2x - 3y + 4z)$.

Solution. Compute the following expressions:

1. $4x + 3x = 7x$.
2. $3x - 2x + 5x = 6x$.
3. $3x - 2(1-x) + 5x = 3x - 2 + 2x + 5x = 10x - 2$.
4. $(3x + 4y) + (2x - y) = 3x + 2x + 4y - y = 5x + 3y$.
5.
$$\begin{aligned}&3(x-y) + 4(y-x) + 5(2x-y) \\ &= 3x - 3y + 4y - 4x + 10x - 5y \\ &= 3x - 4x + 10x - 3y + 4y - 5y \\ &= 9x - 4y.\end{aligned}$$

6.
$$\begin{aligned}
2(x+y) + 3(y-x) + 4y + 3x &= 2x + 2y + 3y - 3x + 4y + 3x \\
&= 2x - 3x + 3x + 2y + 3y + 4y \\
&= 2x + 9y.
\end{aligned}$$

7.
$$\begin{aligned}
2(x+y+z) + 3(2x-y-z) &= 2x + 2y + 2z + 6x - 3y - 3z \\
&= 2x + 6x + 2y - 3y + 2z - 3z \\
&= 8x - y - z.
\end{aligned}$$

8.
$$\begin{aligned}
&4(x-y-z) + 3(z-y+x) - 2(2x-3y+4z) \\
&= 4x - 4y - 4z + 3z - 3y + 3x - 4x + 6y - 8z \\
&= 4x + 3x - 4x - 4y - 3y + 6y - 4z + 3z - 8z \\
&= 3x - y - 9z.
\end{aligned}$$

Problem 2.

Given that $A = x^4 + x^3 + x^2 + x + 1$, $B = 2x^4 + 3x^3 + 5x + 6$ and $C = 3x^3 + 5x - 12$.
Compute

1. $A + B$;
2. $B + C$;
3. $C + A$;
4. $A + B + C$;
5. $A - B + C$;
6. $A + B - C$.

Solution. Compute

1. $A + B$
   We have
   $$\begin{aligned}
   A + B &= x^4 + x^3 + x^2 + x + 1 + 2x^4 + 3x^3 + 5x + 6 \\
   &= x^4 + 2x^4 + x^3 + 3x^3 + x^2 + x + 5x + 1 + 6 \\
   &= 3x^4 + 4x^3 + x^2 + 6x + 7.
   \end{aligned}$$

2.2. Polynomials

2. $B + C$
   We have
   $$B + C = 2x^4 + 3x^3 + 5x + 6 + 3x^3 + 5x - 12$$
   $$= 2x^4 + 3x^3 + 3x^3 + 5x + 5x + 6 - 12$$
   $$= 2x^4 + 6x^3 + 10x - 6.$$

3. $C + A$
   We have
   $$C + A = 3x^3 + 5x - 12 + x^4 + x^3 + x^2 + x + 1$$
   $$= x^4 + 3x^3 + x^3 + x^2 + 5x + x - 12 + 1$$
   $$= x^4 + 4x^3 + x^2 + 6x - 11.$$

4. $A + B + C$
   We have
   $$A + B + C$$
   $$= x^4 + x^3 + x^2 + x + 1 + 2x^4 + 3x^3 + 5x + 6 + 3x^3 + 5x - 12$$
   $$= x^4 + 2x^4 + x^3 + 3x^3 + 3x^3 + x^2 + x + 5x + 5x + 1 + 6 - 12$$
   $$= 3x^4 + 7x^3 + x^2 + 11x - 5.$$

5. $A - B + C$
   We have
   $$A - B + C$$
   $$= \left(x^4 + x^3 + x^2 + x + 1\right) - \left(2x^4 + 3x^3 + 5x + 6\right)$$
   $$+ \left(3x^3 + 5x - 12\right)$$
   $$= x^4 + x^3 + x^2 + x + 1 - 2x^4 - 3x^3 - 5x - 6 + 3x^3 + 5x - 12$$
   $$= x^4 - 2x^4 + x^3 - 3x^3 + 3x^3 + x^2 + x - 5x + 5x + 1 - 6 - 12$$
   $$= -x^4 + x^3 + x^2 + x - 17.$$

6. $A + B - C$
   We have
   $$A + B - C$$
   $$= \left(x^4 + x^3 + x^2 + x + 1\right) + \left(2x^4 + 3x^3 + 5x + 6\right)$$

$$-\left(3x^3+5x-12\right)$$
$$=x^4+x^3+x^2+x+1+2x^4+3x^3+5x+6-3x^3-5x+12$$
$$=x^4+2x^4+x^3+3x^3-3x^3+x^2+x+5x-5x+1+6+12$$
$$=3x^4+x^3+x^2+x+19.$$

Problem 3. Expand the following expressions:

1. $(x+3)(x+4)$;
2. $(x+5)(x-6)$;
3. $(x+1)(x+2)(x-3)$;
4. $(x+y)(2x-y)$;
5. $x(x+1)(x+2)(x+8)$.

Solution. Expand the following expressions:

1. $(x+3)(x+4)$
   We have
   $$(x+3)(x+4)=x^2+4x+3x+12$$
   $$=x^2+7x+12.$$

2. $(x+5)(x-6)$
   We have
   $$(x+5)(x-6)=x^2-6x+5x-30$$
   $$=x^2-x-30.$$

3. $(x+1)(x+2)(x-3)$
   We have
   $$(x+1)(x+2)(x-3)=\left(x^2+2x+x+2\right)(x-3)$$
   $$=\left(x^2+3x+2\right)(x-3)$$
   $$=x^3-3x^2+3x^2-9x+2x-6$$
   $$=x^3-7x-6.$$

2.2. Polynomials

4. $(x+y)(2x-y)$
   We have
   $$(x+y)(2x-y) = 2x^2 - xy + 2xy - y^2$$
   $$= 2x^2 + xy - y^2.$$

5. $x(x+1)(x+2)(x+8)$
   We have
   $$x(x+1)(x+2)(x+8) = (x^2+x)(x^2+8x+2x+16)$$
   $$= (x^2+x)(x^2+10x+16)$$
   $$= x^4 + 10x^3 + 16x^2 + x^3 + 10x^2 + 16x$$
   $$= x^4 + 11x^3 + 26x^2 + 16x.$$

Problem 4.

Calculate the following expressions:

1. $(x+1)(x+2) + (x+2)(x+3)$;
2. $(x+1)(x^2+2) + (x^2+1)(x-1)$;
3. $(x+5)(x-1) + (x+1)(x-10)$;
4. $5(x+1)(x-3) + 7x(3-x)$;
5. $(1-x)(2-x)(3-x) + (4-x)(5-x)(6-x)$.

Solution.

Calculate the following expressions:

1. $(x+1)(x+2) + (x+2)(x+3)$
   We have
   $$(x+1)(x+2) + (x+2)(x+3)$$
   $$= x^2 + 2x + x + 2 + x^2 + 3x + 2x + 6$$
   $$= 2x^2 + 8x + 8.$$

2. $(x+1)(x^2+2) + (x^2+1)(x-1)$
   We have
   $$(x+1)(x^2+2) + (x^2+1)(x-1)$$

$$= x^3 + 2x + x^2 + 2 + x^3 - x^2 + x - 1$$
$$= x^3 + x^3 + x^2 - x^2 + 2x + x + 2 - 1$$
$$= 2x^3 + 3x + 1.$$

3. $(x+5)(x-1) + (x+1)(x-10)$
   We have

$$(x+5)(x-1) + (x+1)(x-10)$$
$$= x^2 - x + 5x - 5 + x^2 - 10x + x - 10$$
$$= 2x^2 - 5x - 15.$$

4. $5(x+1)(x-3) + 7x(3-x)$
   We have

$$5(x+1)(x-3) + 7x(3-x)$$
$$= 5(x^2 - 3x + x - 3) + 21x - 7x^2$$
$$= 5(x^2 - 2x - 3) + 21x - 7x^2$$
$$= 5x^2 - 10x - 15 + 21x - 7x^2$$
$$= -2x^2 + 11x - 15.$$

5. $(1-x)(2-x)(3-x) + (4-x)(5-x)(6-x)$
   We have

$$(1-x)(2-x)(3-x) + (4-x)(5-x)(6-x)$$
$$= (2 - x - 2x + x^2)(3-x) + (20 - 4x - 5x + x^2)(6-x)$$
$$= (2 - 3x + x^2)(3-x) + (20 - 9x + x^2)(6-x)$$
$$= 6 - 2x - 9x + 3x^2 + 3x^2 - x^3 + 120 - 20x - 54x + 9x^2 + 6x^2$$
$$- x^3$$
$$= -2x^3 + 21x^2 - 85x + 126.$$

Chapter 3

Methods in Factorization

Factorization is a critical technique in algebra that involves expressing a mathematical expression as a product of its factors. These factors are usually simpler components, such as numbers, variables, or polynomials, which, when multiplied together, yield the original expression. Factorization is essential because it simplifies complex expressions, making them easier to work with, solve, and understand.

3.1 Common Factor Methods

This method involves identifying and extracting the greatest common factor (GCF) from the terms of the expression.

Example 25

Factor the following expressions:

1. $a + ab$;
2. $a^2 + a$;
3. $ab + ac + ad$;
4. $2x + 2y + 2z$;
5. $ax + bx - 2x$.

Chapter 3. Methods in Factorization

Solution. Factor the following expressions:

1. $a + ab$
   We have $a + ab = a(1 + b)$.

   Remark 5. a is called the common factor of a and ab. That is, to factor an expression, we have to factor the common factor of each terms.

2. $a^2 + a$
   We have $a^2 + a = a(a + 1)$.

3. $ab + ac + ad$
   We have $ab + ac + ad = a(b + c + d)$.

4. $2x + 2y + 2z$
   We have $2x + 2y + 2z = 2(x + y + z)$.

5. $ax + bx - 2x$
   We have $ax + bx - 2x = x(a + b - 2)$.

> **Example 26**
>
> Factor the following expressions:
>
> 1. $3x - 3y + 6z$;
> 2. $x(x - 2) - 3(x - 2)$;
> 3. $ax + ay + bx + by$;
> 4. $x^2 + 2x + 3x + 6$;
> 5. $xyz + abxy$.

Solution. Factor the following expressions:

1. $3x - 3y + 6z$
   We have $3x - 3y + 6z = 3(x - y + 2z)$.

2. $x(x - 2) - 3(x - 2)$
   We have $x(x - 2) - 3(x - 2) = (x - 2)(x - 3)$.

3. $ax + ay + bx + by$

   We have $ax + ay + bx + by = a(x+y) + b(x+y) = (x+y)(a+b)$.

4. $x^2 + 2x + 3x + 6$
We have $x^2 + 2x + 3x + 6 = x(x+2) + 3(x+2) = (x+2)(x+3)$.

5. $xyz + abxy$
We have $xyz + abxy = xy(z + ab)$.

3.2 Splitting Middle Terms

A quadratic is a polynomial in form $ax^2 + bx + c$, where a, b, c are real numbers and $a \neq 0$.
To factor $ax^2 + bx + c$, we have to write it in form
$$ax^2 + bx + c = (a_1x + b_1)(a_2x + b_2).$$

Observe that
$$\begin{aligned}ax^2 + bx + c &= (a_1x + b_1)(a_2x + b_2)\\ &= a_1a_2x^2 + a_1b_2x + a_2b_1x + b_1b_2\\ &= a_1a_2x^2 + (a_1b_2 + a_2b_1)x + b_1b_2.\end{aligned}$$

We obtain $\begin{cases} a_1a_2 = a \\ a_1b_2 + a_2b_1 = b \\ b_1b_2 = c \end{cases}$.

Then
$$\begin{cases}(a_1b_2)(a_2b_1) = ac\\ a_1b_2 + a_2b_1 = b\end{cases}$$
or
$$\begin{cases}xy = ac\\ x + y = b\end{cases}.$$

Thus, to factor a quadratic, we have to find two numbers x and y such that their product and their sum are ac and b respectively. Then we replace b by $x + y$ in $ax^2 + bx + c$. The following examples will illustrate the readers step by step in factoring quadratics.

> **Example 27**
> Factor the following expressions:
> 1. $x^2 - 4x + 3$;
> 2. $x^2 - 5x + 6$;

3. $x^2 - 7x + 10$;

4. $x^2 + 8x + 7$;

5. $x^2 - 12x + 20$.

Solution. Factor the following expressions:

1. $x^2 - 4x + 3$

   Obseve that $a = 1$, $b = -4$ and $c = 3$. Then $ac = 3$. We have to find two numbers x and y such that $xy = 3$ and $x+y = -4$. To make it easier, we can list all pair of numbers such that $xy = 3$ and find the pair that satisfies $x + y = -4$.
   See the list below:
   $$\begin{array}{cc} 1 & 3 \\ -1 & -3 \end{array}$$

   We see that -1 and -3 are the pair that satisfies the condition.
   We obtain
   $$\begin{aligned} x^2 - 4x + 3 &= x^2 - x - 3x + 3 \\ &= x(x - 1) - 3(x - 1) \\ &= (x - 1)(x - 3). \end{aligned}$$

2. $x^2 - 5x + 6$
   We have
   $$\begin{aligned} x^2 - 5x + 6 &= x^2 - 2x - 3x + 6 \\ &= x(x - 2) - 3(x - 2) \\ &= (x - 2)(x - 3). \end{aligned}$$

3. $x^2 - 7x + 10$
   We have
   $$\begin{aligned} x^2 - 7x + 10 &= x^2 - 2x - 5x + 10 \\ &= x(x - 2) - 5(x - 2) \\ &= (x - 2)(x - 5). \end{aligned}$$

3.2. Splitting Middle Terms

4. $x^2 + 8x + 7$
   We have
   $$x^2 + 8x + 7 = x^2 + x + 7x + 7$$
   $$= x(x+1) + 7(x+1)$$
   $$= (x+1)(x+7).$$

5. $x^2 - 12x + 20$
   We have
   $$x^2 - 12x + 20 = x^2 - 10x - 2x + 20$$
   $$= x(x-10) - 2(x-10)$$
   $$= (x-10)(x-2).$$

> **Example 28**
> Factor the following expressions:
>
> 1. $3a^2b - 18ab^2$;
> 2. $x^2 - 11x + 24$;
> 3. $x(x-4) - 5$;
> 4. $6x^2 + 13x - 8$;
> 5. $(a-3)^2 - (a-3)$;
> 6. $x^2 - x - y^2 - y$.

Solution. Factor the following expressions:

1. $3a^2b - 18ab^2$
   We have $3a^2b - 18ab^2 = 3ab(a - 6b)$.

2. $x^2 - 11x + 24$
   We have
   $$x^2 - 11x + 24 = x^2 - 3x - 8x + 24$$
   $$= x(x-3) - 8(x-3)$$
   $$= (x-3)(x-8).$$

3. $x(x-4) - 5$

   We have
   $$x(x-4) - 5 = x^2 - 4x - 5$$

$$\begin{aligned} &= x^2 + x - 5x - 5 \\ &= x(x+1) - 5(x+1) \\ &= (x+1)(x-5). \end{aligned}$$

4. $6x^2 + 13x - 8$

   We have
   $$\begin{aligned} 6x^2 + 13x - 8 &= 6x^2 + 16x - 3x - 8 \\ &= 2x(3x+8) - (3x+8) \\ &= (3x+8)(2x-1). \end{aligned}$$

5. $(a-3)^2 - (a-3)$

   We have
   $$\begin{aligned} (a-3)^2 - (a-3) &= (a-3)(a-3-1) \\ &= (a-3)(a-4). \end{aligned}$$

6. $x^2 - x - y^2 - y$

   We have
   $$\begin{aligned} x^2 - x - y^2 - y &= x^2 - y^2 - x - y \\ &= (x-y)(x+y) - (x+y) \\ &= (x+y)(x-y-1). \end{aligned}$$

> **Example 29**
>
> Factor the following expressions:
>
> 1. $(x-y)(x-y+5) + 6$;
> 2. $a^2 - 2ab + b^2 - 9$;
> 3. $x^2 - (4a - 3b)x - 12ab$;
> 4. $4(x-3y)^2 - 9(x-3y) + 5$;
> 5. $x^2 + xy - 6y^2 + 5x + 35y - 36$.

Solution. Factor the following expressions:

3.2. Splitting Middle Terms

1. $(x-y)(x-y+5)+6$

    Let $t = x - y$. We obtain
    $$\begin{aligned}(x-y)(x-y+5)+6 &= t(t+5)+6 \\ &= t^2 + 5t + 6 \\ &= t^2 + 2t + 3t + 6 \\ &= t(t+2) + 3(t+2) \\ &= (t+2)(t+3) \\ &= (x-y+2)(x-y+3).\end{aligned}$$

2. $a^2 - 2ab + b^2 - 9$
    We have
    $$\begin{aligned}a^2 - 2ab + b^2 - 9 &= (a-b)^2 - 3^2 \\ &= (a-b-3)(a-b+3).\end{aligned}$$

3. $x^2 - (4a - 3b)x - 12ab$
    We have
    $$\begin{aligned}x^2 - (4a-3b)x - 12ab &= x^2 - 4ax + 3bx - 12ab \\ &= x(x-4a) + 3b(x-4a) \\ &= (x-4a)(x+3b).\end{aligned}$$

4. $4(x-3y)^2 - 9(x-3y) + 5$
    Let $t = x - 3y$. We obtain
    $$\begin{aligned}4(x-3y)^2 - 9(x-3y) + 5 &= 4t^2 - 9t + 5 \\ &= 4t^2 - 4t - 5t + 5 \\ &= 4t(t-1) - 5(t-1) \\ &= (t-1)(4t-5) \\ &= (x-3y-1)[4(x-3y)-5] \\ &= (x-3y-1)(4x-12y-5).\end{aligned}$$

5. $x^2 + xy - 6y^2 + 5x + 35y - 36$
    We have
    $$x^2 + xy - 6y^2 + 5x + 35y - 36$$

$$= x^2 + xy + 5x - 6y^2 + 35y - 36$$
$$= x^2 + (y+5)x - (6y^2 - 35y + 36)$$
$$= x^2 + (y+5)x - (6y^2 - 8y - 27y + 36)$$
$$= x^2 + (y+5)x - [2y(3y-4) - 9(3y-4)]$$
$$= x^2 + (y+5)x - (3y-4)(2y-9)$$
$$= x^2 + (3y-4)x - (2y-9)x - (3y-4)(2y-9)$$
$$= x(x+3y-4) - (2y-9)(x+3y-4)$$
$$= (x+3y-4)(x-2y+9).$$

Exercises

Problem 1. Factor the following expressions:

1. $5x - 10$;
2. $7x - 21$;
3. $-3x + 9$;
4. $-9x + 18$;
5. $3x + 6$;
6. $5x^2 - 20x$;
7. $6x^2 - 9x$;
8. $7x^2 + 14x$;
9. $13x^2 - 26x$;
10. $5ax + ab$;
11. $10a^2b - 7ab^2$;
12. $2a + 2ax$;
13. $a^2b^2c^2 - 2ab^2c^3$;
14. $ax + ay + az$;
15. $2ax + 3ay - 4az$;
16. $2ax - 6ay + 4az$;
17. $7abc + 8abx - 9abcx$;
18. $5a^2xy + 10xy$;
19. $7a^2x^3y^3 - 14x^4y^4$;
20. $4x^2yz - 6x^2y^2z^2 + 8xyz$.

Problem 2. Factor the following expressions:

1. $x(x+1) + 2(x+1)$;
2. $3x(x-1) - 4(x-1)$;
3. $5x(x+2) - 2(x+2)$;
4. $7x(x-1) - 14(x-1)$;
5. $2x(1-x) + 3(x-1)$;
6. $-2x(2-3x) + 3(3x-2)$;

7. $7x(2x-1) - 8(1-2x)$;

8. $2x(1-x) + 3(x-1)$;

9. $3x(x-1) + (2x-3) - x(2x-3)$;

10. $x(x+2) - 2(2x-1) - x(2x-1)$.

Problem 3. Factor the following expressions:

1. $ax + ay - bx - by$;

2. $ax - 2bx + ay - 2by$;

3. $4a - 4x + a^2 - x^2$;

4. $-2a + 2x + 4(a-x)(a+3b)$;

5. $3a - 6b + (a-2b)(x+y)$;

6. $ax + bx + cx - (a+b) - (b+c) + b$;

7. $2x - 4y + ax - 2ay$;

8. $a^2b - ab^2 - ax + bx$.

Problem 4. Factor the following expressions:

1. $(x+y)^2 - (2x-y)^2$

2. $(2x+3y)^2 - (4x-2y)^2$

3. $(x+y)^2 - 4z^2$

4. $(x+y)^2 - (z+t)^2$

5. $9(x-y)^2 - 4(z+t)^2$

6. $(x^2 + 2ax + a^2) - 16(y-z)^2$.

Problem 5. Find the value of a such that the following equalities hold.

1. $(3x-5y)(3x+5y) = 9x^2 - 2ay^2$;

2. $(4x-y)(4x+y) = 16x^2 - 5ay^2$;

3. $(x-1)(x+1)(x-2) = x^3 - 2x^2 - x + a$;

4. $(x-2)(x+2)(x-3) = x^3 + ax^2 - 4x + 12$.

Problem 6. Factor the following expressions:

3.2. Splitting Middle Terms

1. $x^2 - 4$;
2. $9x^2 - 1$;
3. $4x^2 - 121$;
4. $4x^2 - 49$;
5. $2x^2 - 8$;
6. $7x^2 - 28$;
7. $x^2 + 2xy + y^2 - 4$;
8. $4x^2 - 4xy + y^2 - 9$;
9. $5x^2y^2 - 125$;
10. $2xy^2 - 8x$;
11. $5xy^2 - 5x^2 + 2x - 2y$;
12. $4ax^2 - 4ax + axy^2 - ay^2$;
13. $9x^2 - 4(2x - 1)^2$
14. $3x - 4x^2 + (3 - 4x)^2$
15. $4xy - 5xy^2 + 2(4 - 5y)^2$
16. $75y^2 - 3(x - 1)^2$

Problem 7. Factor the following expressions:

1. $9(x - 1)^2 - 4(2x + 3)^2$;
2. $4(y + 2)^2 - y^2$;
3. $3(x + 1) - x^2 - 2x - 1$;
4. $(2x - 5)^2 - 4x^2 + 25$;
5. $16(x + 3)^2 - (x - 1)^2$;
6. $x^3 - x$;
7. $t^4 - 1$;
8. $4x^2 - 16x$;
9. $(x + 1)^2 - 2(x^2 - 1) + 2x + 2$;
10. $x^5 - x$;
11. $x(y - 1)^2 - 4x$.

Problem 8. Find the value of a such that
$$3x^3 + 2x^2 + 1 = (x + 1)\left(3x^2 - ax + 1\right).$$

Problem 9. Factor the following expressions:

1. $x^2 + 3x + 2$;
2. $x^2 - 3x + 2$;
3. $x^2 - 5x + 4$;
4. $x^2 - 11x + 28$;
5. $x^2 - 7x + 10$;
6. $x^2 - 9x + 18$;
7. $x^2 - 7x + 12$;
8. $x^2 - 12x + 20$;
9. $x^2 - 8x + 16$;
10. $x^2 - 11x + 30$;
11. $2x^2 - 12x + 10$;
12. $2x^2 - 3x + 1$;
13. $4x^2 - 5x + 1$;
14. $8x^2 - x - 7$;
15. $6x^2 - 2x - 4$;
16. $7x^2 - 4x - 3$;
17. $3x^2 - 5x + 2$;
18. $6x^2 - x - 5$;
19. $10x^2 - x - 9$;
20. $8x^2 - 4x - 4$.

Problem 10. Calculate the following expressions:

1. $A = 4x - \dfrac{x-1}{x-2}$;
2. $B = x - 1 + \dfrac{3x}{x-1}$;
3. $C = 3x - 2 + \dfrac{2}{x-2}$;
4. $D = 4x - 1 - \dfrac{x-1}{2x-1}$;
5. $E = 3x - 5 - \dfrac{3x-1}{x-1}$;
6. $F = 2x - 2 + \dfrac{3x-1}{x+2}$;
7. $G = -3x + 1 - \dfrac{1}{x-1}$;
8. $H = 4x - 2 - \dfrac{3x-1}{x-2}$;
9. $I = 4x - 3 - \dfrac{2x-1}{x-2}$;

3.2. Splitting Middle Terms

10. $J = -7x + 1 + \dfrac{7x-1}{3x-1}$.

Problem 11. Find the value of a, b and c such that

1. $\dfrac{3x^2 - 2x + 1}{x - 1} = ax + b + \dfrac{c}{x - 1}$;

2. $\dfrac{5x^2 - x - 1}{x - 1} = ax + b + \dfrac{c}{x - 1}$;

3. $\dfrac{x^2 - 2x + 4}{x + 2} = ax + b + \dfrac{c}{x + 2}$;

4. $\dfrac{x^2 - 6x + 3}{x + 2} = ax + b + \dfrac{c}{x + 2}$.

Problem 12. Compute the following expressions:

1. $\dfrac{4}{x - 1} - \dfrac{2}{x^2 - 1}$;

2. $\dfrac{7x}{x + 1} - \dfrac{3x}{x^2 - 1}$;

3. $\dfrac{7x - 1}{x^2 - x} - \dfrac{3x - 1}{x - 1}$;

4. $\dfrac{8x - 1}{x^2 - x} - \dfrac{4x - 1}{x} + \dfrac{3}{x - 1}$;

5. $\dfrac{6x - 1}{x^3 - x} - \dfrac{3x - 1}{x} + \dfrac{7}{x - 1}$;

6. $\dfrac{7x - 3}{x^3 - x} + \dfrac{4x - 1}{x + 1} - \dfrac{5x - 1}{x - 1}$;

7. $\dfrac{7}{x^4 + x^2 + 1} - \dfrac{8x - 1}{x^2 - x + 1}$;

8. $\dfrac{9x - 1}{x^4 + x^2 + 1} + \dfrac{8x - 1}{x^2 - x + 1} + \dfrac{7x - 1}{x^2 + x + 1}$.

Problem 13. Compute the following expressions:

Chapter 3. Methods in Factorization

1. $A = \dfrac{\dfrac{1}{x} + \dfrac{1}{x-1}}{\dfrac{2x+1}{x+1} + \dfrac{3}{x-1}}$;

2. $B = \dfrac{\dfrac{3x-1}{x^2-1} + \dfrac{4x-1}{x-1}}{\dfrac{3x-1}{x+1} - \dfrac{4}{x^2-1}}$;

3. $C = \dfrac{\dfrac{1}{x} + \dfrac{1}{x-1} + \dfrac{1}{x^2-x}}{\dfrac{2x-1}{x^2-x} + \dfrac{3}{x}}$;

4. $D = \dfrac{\dfrac{1}{x} + \dfrac{1}{x+1} + \dfrac{3}{x-1}}{\dfrac{4x-1}{x^2-x} + \dfrac{3x-1}{x^2-1}}$;

5. $E = \dfrac{1}{x + \dfrac{1}{x + \dfrac{1}{x}}}$.

Problem 14. Find the value of $A = \dfrac{2x^3 - 8x}{4x^3 - 16x^2 + 16x}$ for $x = 2024$.

Solutions

Problem 1. Factor the following expressions:

1. $5x - 10$;
2. $7x - 21$;
3. $-3x + 9$;
4. $-9x + 18$;
5. $3x + 6$;
6. $5x^2 - 20x$;
7. $6x^2 - 9x$;
8. $7x^2 + 14x$;
9. $13x^2 - 26x$;
10. $5ax + ab$;
11. $10a^2b - 7ab^2$;
12. $2a + 2ax$;
13. $a^2b^2c^2 - 2ab^2c^3$;
14. $ax + ay + az$;
15. $2ax + 3ay - 4az$;
16. $2ax - 6ay + 4az$;
17. $7abc + 8abx - 9abcx$;
18. $5a^2xy + 10xy$;
19. $7a^2x^3y^3 - 14x^4y^4$;
20. $4x^2yz - 6x^2y^2z^2 + 8xyz$.

Solution. Factor the following expressions:

1. $5x - 10 = 5(x - 2)$
2. $7x - 21 = 7(x - 3)$
3. $-3x + 9 = -3(x - 3)$
4. $-9x + 18 = -9(x - 2)$
5. $3x + 6 = 3(x + 2)$
6. $5x^2 - 20x = 5x(x - 4)$

Chapter 3. Methods in Factorization

7. $6x^2 - 9x = 3x(2x - 3)$
8. $7x^2 + 14x = 7x(x + 2)$
9. $13x^2 - 26x = 13x(x - 2)$
10. $5ax + ab = a(5x + b)$
11. $10a^2b - 7ab^2 = ab(10a - 7b)$
12. $2a + 2ax = 2a(1 + x)$
13. $a^2b^2c^2 - 2ab^2c^3 = ab^2c^2(a - 2c)$
14. $ax + ay + az = a(x + y + z)$
15. $2ax + 3ay - 4az = a(2x + 3y - 4z)$
16. $2ax - 6ay + 4az = 2a(x - 3y + 2z)$
17. $7abc + 8abx - 9abcx = ab(7c + 8x - 9cx)$
18. $5a^2xy + 10xy = 5xy(a^2 + 2)$
19. $7a^2x^3y^3 - 14x^4y^4 = 7x^3y^3(a^2 - 2xy)$
20. $4x^2yz - 6x^2y^2z^2 + 8xyz = 2xyz(2x - 3xyz + 4)$

Problem 2. Factor the following expressions:

1. $x(x + 1) + 2(x + 1)$;
2. $3x(x - 1) - 4(x - 1)$;
3. $5x(x + 2) - 2(x + 2)$;
4. $7x(x - 1) - 14(x - 1)$;
5. $2x(1 - x) + 3(x - 1)$;
6. $-2x(2 - 3x) + 3(3x - 2)$;
7. $7x(2x - 1) - 8(1 - 2x)$;
8. $2x(1 + x) + 3(-x - 1)$;
9. $3x(x - 1) + (2x - 3) - x(2x - 3)$;
10. $x(x + 2) - 2(2x - 1) - x(2x - 1)$.

3.2. Splitting Middle Terms

Solution. Factor the following expressions:
1. $x(x+1) + 2(x+1) = (x+1)(x+2)$
2. $3x(x-1) - 4(x-1) = (x-1)(3x-4)$
3. $5x(x+2) - 2(x+2) = (x+2)(5x-2)$
4. $7x(x-1) - 14(x-1) = (x-1)(7x-14) = 7(x-1)(x-2)$
5.
$$2x(1-x) + 3(x-1) = -2x(x-1) + 3(x-1)$$
$$= (x-1)(-2x+3)$$

6. We have
$$-2x(2-3x) + 3(3x-2) = 2x(3x-2) + 3(3x-2)$$
$$= (3x-2)(2x+3)$$
Therefore, $-2x(2-3x) + 3(3x-2) = (3x-2)(2x+3)$.

7. We have
$$7x(2x-1) - 8(1-2x) = 7x(2x-1) + 8(2x-1)$$
$$= (2x-1)(7x+8).$$
Therefore, $7x(2x-1) - 8(1-2x) = (2x-1)(7x+8)$.

8. We have
$$2x(1+x) + 3(-x-1) = 2x(x+1) - 3(x+1)$$
$$= (x+1)(2x-3).$$
Therefore, $2x(1+x) + 3(-x-1) = (x+1)(2x-3)$.

9. We have
$$3x(x-1) + (2x-3) - x(2x-3)$$
$$= 3x(x-1) + (2x-3)(1-x)$$
$$= 3x(x-1) - (2x-3)(x-1)$$
$$= (x-1)[3x - (2x-3)]$$
$$= (x-1)(3x - 2x + 3)$$
$$= (x-1)(x+3)$$
Therefore, $3x(x-1) + (2x-3) - x(2x-3) = (x-1)(x+3)$.

Chapter 3. Methods in Factorization

10. We have
$$x(x+2) - 2(2x-1) - x(2x-1)$$
$$= x(x+2) - (2x-1)(2+x)$$
$$= (x+2)[x - (2x-1)]$$
$$= (x+2)(x - 2x + 1)$$
$$= (x+2)(-x+1).$$

Therefore, $x(x+2) - 2(2x-1) - x(2x-1) = (x+2)(-x+1)$.

Problem 3. Factor the following expressions:

1. $ax + ay - bx - by$;
2. $ax - 2bx + ay - 2by$;
3. $4a - 4x + a^2 - x^2$;
4. $-2a + 2x + 4(a-x)(a+3b)$;
5. $3a - 6b + (a-2b)(x+y)$;
6. $ax + bx + cx - (a+b) - (b+c) + b$;
7. $2x - 4y + ax - 2ay$;
8. $a^2 b - ab^2 - ax + bx$.

Solution. Factor the following expressions:

1. $ax + ay - bx - by$
   We have
   $$ax + ay - bx - by = a(x+y) - b(x+y)$$
   $$= (x+y)(a-b)$$
   Therefore, $ax + ay - bx - by = (x+y)(a-b)$.

2. $ax - 2bx + ay - 2by$
   We have
   $$ax - 2bx + ay - 2by = x(a-2b) + y(a-2b)$$
   $$= (a-2b)(x+y).$$
   Therefore, $ax - 2bx + ay - 2by = (a-2b)(x+y)$.

3.2. Splitting Middle Terms

3. $4a - 4x + a^2 - x^2$
   We have
   $$4a - 4x + a^2 - x^2 = 4(a - x) + (a - x)(a + x)$$
   $$= (a - x)(4 + a + x).$$
   Therefore, $4a - 4x + a^2 - x^2 = (a - x)(4 + a + x)$.

4. $-2a + 2x + 4(a - x)(a + 3b)$
   We have
   $$-2a + 2x + 4(a - x)(a + 3b)$$
   $$= -2(a - x) + 4(a - x)(a + 3b)$$
   $$= (a - x)[-2 + 4(a + 3b)]$$
   $$= (a - x)(-2 + 4a + 12b)$$
   $$= (a - x)(4a + 12b - 2)$$
   $$= 2(a - x)(2a + 6b - 1).$$
   Therefore, $-2a + 2x + 4(a - x)(a + 3b) = 2(a - x)(2a + 6b - 1)$.

5. $3a - 6b + (a - 2b)(x + y)$
   We have
   $$3a - 6b + (a - 2b)(x + y)$$
   $$= 3(a - 2b) + (a - 2b)(x + y)$$
   $$= (a - 2b)(3 + x + y)$$
   $$= (a - 2b)(x + y + 3).$$
   Therefore, $3a - 6b + (a - 2b)(x + y) = (a - 2b)(x + y + 3)$.

6. $ax + bx + cx - (a + b) - (b + c) + b$
   We have
   $$ax + bx + cx - (a + b) - (b + c) + b$$
   $$= x(a + b + c) - a - b - b - c + b$$
   $$= x(a + b + c) - a - b - c$$
   $$= x(a + b + c) - (a + b + c)$$
   $$= (x - a)(a + b + c).$$
   Therefore, $ax + bx + cx - (a + b) - (b + c) + b = (x - a)(a + b + c)$.

Chapter 3. Methods in Factorization

7. $2x - 4y + ax - 2ay$
We have
$$2x - 4y + ax - 2ay = 2(x - 2y) + a(x - 2y)$$
$$= (x - 2y)(2 + a)$$
$$= (x - 2y)(a + 2).$$

8. $a^2b - ab^2 - ax + bx$
We have
$$a^2b - ab^2 - ax + bx = ab(a - b) - x(a - b)$$
$$= (a - b)(ab - x).$$

Therefore, $a^2b - ab^2 - ax + bx = (a - b)(ab - x)$.

Problem 4. Factor the following expressions:

1. $(x + y)^2 - (2x - y)^2$
2. $(2x + 3y)^2 - (4x - 2y)^2$
3. $(x + y)^2 - 4z^2$
4. $(x + y)^2 - (z + t)^2$
5. $9(x - y)^2 - 4(z + t)^2$
6. $(x^2 + 2ax + a^2) - 16(y - z)^2$.

Solution. Factor the following expressions:

1. $(x + y)^2 - (2x - y)^2$
We have
$$(x + y)^2 - (2x - y)^2 = [(x + y) - (2x - y)][(x + y) + (2x - y)]$$
$$= (x + y - 2x + y)(x + y + 2x - y)$$
$$= (-x + 2y)(3x)$$
$$= 3x(-x + 2y).$$

Therefore, $(x + y)^2 - (2x - y)^2 = 3x(-x + 2y)$.

66

3.2. Splitting Middle Terms

2. $(2x + 3y)^2 - (4x - 2y)^2$
   We have
   $$(2x + 3y)^2 - (4x - 2y)^2$$
   $$= [(2x + 3y) - (4x - 2y)][(2x + 3y) + (4x - 2y)]$$
   $$= (2x + 3y - 4x + 2y)(2x + 3y + 4x - 2y)$$
   $$= (-2x + 5y)(6x + y).$$

   Therefore, $(2x + 3y)^2 - (4x - 2y)^2 = (-2x + 5y)(6x + y)$.

3. $(x + y)^2 - 4z^2$
   We have
   $$(x + y)^2 - 4z^2 = (x + y)^2 - (2z)^2$$
   $$= [(x + y) - 2z][(x + y) + 2z]$$
   $$= (x + y - 2z)(x + y + 2z).$$

   Therefore, $(x + y)^2 - 4z^2 = (x + y - 2z)(x + y + 2z)$.

4. $(x + y)^2 - (z + t)^2$
   We have
   $$(x + y)^2 - (z + t)^2$$
   $$= [(x + y) - (z + t)][(x + y) + (z + t)]$$
   $$= (x + y - z - t)(x + y + z + t).$$

   Therefore, $(x + y)^2 - (z + t)^2 = (x + y - z - t)(x + y + z + t)$.

5. $9(x - y)^2 - 4(z + t)^2$
   We have
   $$9(x - y)^2 - 4(z + t)^2$$
   $$= [3(x - y)]^2 - [2(z + t)]^2$$
   $$= [3(x - y) - 2(z + t)][3(x - y) + 2(z + t)]$$
   $$= (3x - 3y - 2z - 2t)(3x - 3y + 2z + 2t).$$

   Therefore,
   $$9(x - y)^2 - 4(z + t)^2$$
   $$= (3x - 3y - 2z - 2t)(3x - 3y + 2z + 2t).$$

6. $(x^2 + 2ax + a^2) - 16(y-z)^2$.
We have

$$(x^2 + 2ax + a^2) - 16(y-z)^2$$
$$= (x+a)^2 - [4(y-z)]^2$$
$$= [(x+a) - 4(y-z)][(x+a) + 4(y-z)]$$
$$= (x+a-4y+4z)(x+a+4y-4z).$$

Therefore,

$$(x^2 + 2ax + a^2) - 16(y-z)^2$$
$$= (x+a-4y+4z)(x+a+4y-4z).$$

Problem 5. Find the value of a such that the following equalities hold.

1. $(3x-5y)(3x+5y) = 9x^2 - 2ay^2$;
2. $(4x-y)(4x+y) = 16x^2 - 5ay^2$;
3. $(x-1)(x+1)(x-2) = x^3 + 2x^2 - x + a$;
4. $(x-2)(x+2)(x-3) = x^3 + ax^2 - 4x + 12$.

Solution. Find the value of a such that the following equalities hold.

1. $(3x-5y)(3x+5y) = 9x^2 - 2ay^2$
We have

$$(3x-5y)(3x+5y) = 9x^2 - 2ay^2$$
$$(3x)^2 - (5y)^2 = 9x^2 - 2ay^2$$
$$9x^2 - 25y^2 = 9x^2 - 2ay^2.$$

Comparing the coefficient of y^2, we obtain

$$-2a = -25.$$

It follows that $a = \dfrac{25}{2}$.

Therefore, $a = \dfrac{25}{2}$.

68

3.2. Splitting Middle Terms

2. $(4x - y)(4x + y) = 16x^2 - 5ay^2$
   We have
   $$(4x - y)(4x + y) = 16x^2 - 5ay^2$$
   $$(4x)^2 - y^2 = 16x^2 - 5ay^2$$
   $$16x^2 - y^2 = 16x^2 - 5ay^2.$$

   Comparing the coefficient of y^2, we obtain
   $$-5a = -1.$$

   It follows that $a = \dfrac{1}{5}$.

   Therefore, $a = \dfrac{1}{5}$.

3. $(x - 1)(x + 1)(x - 2) = x^3 - 2x^2 - x + a$
   We have
   $$(x - 1)(x + 1)(x - 2) = x^3 - 2x^2 - x + a$$
   $$(x^2 - 1)(x - 2) = x^3 - 2x^2 - x + a$$
   $$x^3 - 2x^2 - x + 2 = x^3 - 2x^2 - x + a.$$

   Comparing the constant parts of both polynomials, we obtain
   $$a = 2.$$

   Therefore, $a = 2$.

4. $(x - 2)(x + 2)(x - 3) = x^3 + ax^2 - 4x + 12$
   We have
   $$(x - 2)(x + 2)(x - 3) = x^3 + ax^2 - 4x + 12$$
   $$(x^2 - 4)(x - 3) = x^3 + ax^2 - 4x + 12$$
   $$x^3 - 3x^2 - 4x + 12 = x^3 + ax^2 - 4x + 12.$$

   Comparing the constant parts of both polynomials, we obtain
   $$a = -3.$$

   Therefore, $a = -3$.

Problem 6. Factor the following expressions:

Chapter 3. Methods in Factorization

1. $x^2 - 4$;
2. $9x^2 - 1$;
3. $4x^2 - 121$;
4. $4x^2 - 49$;
5. $2x^2 - 8$;
6. $7x^2 - 28$;
7. $x^2 + 2xy + y^2 - 4$;
8. $4x^2 - 4xy + y^2 - 9$;
9. $5x^2y^2 - 125$;
10. $2xy^2 - 8x$;
11. $5xy^2 - 5x^2 + 2x - 2y$;
12. $4ax^2 - 4ax + xy^2 - ay^2$;
13. $9x^2 - 4(2x - 1)^2$;
14. $3x - 4x^2 + (3 - 4x)^2$;
15. $4xy - 5xy^2 + 2(4 - 5y)^2$;
16. $75y^2 - 3(x - 1)^2$.

Solution. Factor the following expressions:

1. $x^2 - 4$
   We have
   $$x^2 - 4 = x^2 - 2^2$$
   $$= (x - 2)(x + 2).$$
   Therefore, $x^2 - 4 = (x - 2)(x + 2)$.

2. $9x^2 - 1$
   We have
   $$9x^2 - 1 = (3x)^2 - 1^2$$
   $$= (3x - 1)(3x + 1).$$
   Therefore, $9x^2 - 1 = (3x - 1)(3x + 1)$.

3. $4x^2 - 121$
   We have
   $$4x^2 - 121 = (2x)^2 - 11^2$$
   $$= (2x - 11)(2x + 11).$$
   Therefore, $4x^2 - 121 = (2x - 11)(2x + 11)$.

4. $4x^2 - 49$
   We have
   $$4x^2 - 49 = (2x)^2 - 7^2$$

3.2. Splitting Middle Terms

$$= (2x - 7)(2x + 7).$$

Therefore, $4x^2 - 49 = (2x - 7)(2x + 7)$.

5. $2x^2 - 8$
   We have
   $$\begin{aligned} 2x^2 - 8 &= 2\left(x^2 - 4\right) \\ &= 2\left(x^2 - 2^2\right) \\ &= 2(x - 2)(x + 2). \end{aligned}$$

   Therefore, $2x^2 - 8 = 2(x - 2)(x + 2)$.

6. $7x^2 - 28$
   We have
   $$\begin{aligned} 7x^2 - 28 &= 7\left(x^2 - 4\right) \\ &= 7\left(x^2 - 2^2\right) \\ &= 7(x - 2)(x + 2). \end{aligned}$$

   Therefore, $7x^2 - 28 = 7(x - 2)(x + 2)$.

7. $x^2 + 2xy + y^2 - 4$
   We have
   $$\begin{aligned} & x^2 + 2xy + y^2 - 4 \\ &= (x + y)^2 - 2^2 \\ &= (x + y - 2)(x + y + 2). \end{aligned}$$

   Therefore, $x^2 + 2xy + y^2 - 4 = (x + y - 2)(x + y + 2)$.

8. $4x^2 - 4xy + y^2 - 9$
   We have
   $$\begin{aligned} & 4x^2 - 4xy + y^2 - 9 \\ &= (2x - y)^2 - 3^2 \\ &= (2x - y - 3)(2x - y + 3). \end{aligned}$$

   Therefore, $4x^2 - 4xy + y^2 - 9 = (2x - y - 3)(2x - y + 3)$.

9. $5x^2y^2 - 125$
   We have
   $$\begin{aligned}5x^2y^2 - 125 &= 5\left(x^2y^2 - 25\right)\\ &= 5\left[(xy)^2 - 5^2\right]\\ &= 5\left(xy - 5\right)\left(xy + 5\right).\end{aligned}$$
   Therefore, $5x^2y^2 - 125 = 5\left(xy - 5\right)\left(xy + 5\right)$.

10. $2xy^2 - 8x$
    We have
    $$\begin{aligned}2xy^2 - 8x &= 2x\left(y^2 - 4\right)\\ &= 2x\left(y^2 - 2^2\right)\\ &= 2x\left(y - 2\right)\left(y + 2\right).\end{aligned}$$
    Therefore, $2xy^2 - 8x = 2x\left(y - 2\right)\left(y + 2\right)$.

11. $5xy^2 - 5x^2 + 2x - 2y$
    We have
    $$\begin{aligned}5xy^2 - 5x^2 + 2x - 2y &= -5x\left(-y + x\right) + 2\left(x - y\right)\\ &= -5x\left(x - y\right) + 2\left(x - y\right)\\ &= \left(x - y\right)\left(-5x + 2\right).\end{aligned}$$
    Therefore, $5xy^2 - 5x^2 + 2x - 2y = \left(x - y\right)\left(-5x + 2\right)$.

12. $4ax^2 - 4ax + axy^2 - ay^2$
    We have
    $$\begin{aligned}&4ax^2 - 4ax + axy^2 - ay^2\\ &= a\left(4x^2 - 4x + xy^2 - y^2\right)\\ &= a\left[4x\left(x - 1\right) + y^2\left(x - 1\right)\right]\\ &= a\left(x - 1\right)\left(4x + y^2\right).\end{aligned}$$
    Therefore, $4ax^2 - 4ax + axy^2 - ay^2 = a\left(x - 1\right)\left(4x + y^2\right)$.

13. $9x^2 - 4(2x - 1)^2$
    We have
    $$9x^2 - 4(2x - 1)^2 = (3x)^2 - \left[2\left(2x - 1\right)\right]^2$$

3.2. Splitting Middle Terms

$$= [3x - 2(2x-1)][3x + 2(2x-1)]$$
$$= (3x - 4x + 2)(3x + 4x - 2)$$
$$= (-x + 2)(7x - 2).$$

Therefore, $9x^2 - 4(2x-1)^2 = (-x+2)(7x-2)$.

14. $3x - 4x^2 + (3 - 4x)^2$
We have

$$3x - 4x^2 + (3-4x)^2 = x(3-4x) + (3-4x)^2$$
$$= (3-4x)(x + 3 - 4x)$$
$$= (3-4x)(3 - 3x)$$
$$= (4x - 3)(3x - 3)$$
$$= 3(4x - 3)(x - 1).$$

Therefore, $3x - 4x^2 + (3-4x)^2 = 3(4x-3)(x-1)$.

15. $4xy - 5xy^2 + 2(4 - 5y)^2$
We have

$$4xy - 5xy^2 + 2(4-5y)^2 = xy(4 - 5y) + 2(4 - 5y)^2$$
$$= (4 - 5y)[xy + 2(4-5y)]$$
$$= (4 - 5y)(xy + 8 - 10y).$$

Therefore, $4xy - 5xy^2 + 2(4-5y)^2 = (4-5y)(xy + 8 - 10y)$.

16. $75y^2 - 3(x-1)^2$
We have

$$75y^2 - 3(x-1)^2 = 3\left[25y^2 - (x-1)^2\right]$$
$$= 3\left[(5y)^2 - (x-1)^2\right]$$
$$= 3[5y - (x-1)][5y + (x-1)]$$
$$= 3(5y - x + 1)(5y + x - 1).$$

Therefore, $75y^2 - 3(x-1)^2 = 3(5y - x + 1)(5y + x - 1)$.

Problem 7. Factor the following expressions:

1. $9(x-1)^2 - 4(2x+3)^2$;

2. $4(y+2)^2 - y^2$;

3. $3(x+1) - x^2 - 2x - 1$;

4. $(2x-5)^2 - 4x^2 + 25$;

5. $16(x+3)^2 - (x-1)^2$;

6. $x^3 - x$;

7. $t^4 - 1$;

8. $4x^2 - 16x$;

9. $(x+1)^2 - (x^2-1) + 2x + 2$;

10. $x^5 - x$;

11. $x(y-1)^2 - 4x$.

Solution. Factor the following expressions:

1. $9(x-1)^2 - 4(2x+3)^2$
We have
$$9(x-1)^2 - 4(2x+3)^2$$
$$= [3(x-1)]^2 - [2(2x+3)]^2$$
$$= [3(x-1) - 2(2x+3)][3(x-1) + 2(2x+3)]$$
$$= (3x - 3 - 4x - 6)(3x - 3 + 4x + 6)$$
$$= (-x - 9)(7x + 3)$$
$$= -(x+9)(7x+3).$$

Therefore, $9(x-1)^2 - 4(2x+3)^2 = -(x+9)(7x+3)$.

2. $4(y+2)^2 - y^2$
We have
$$4(y+2)^2 - y^2 = [2(y+2)]^2 - y^2$$
$$= [2(y+2) - y][2(y+2) + y]$$
$$= (2y + 4 - y)(2y + 4 + y)$$
$$= (y+4)(3y+4).$$

Therefore, $4(y+2)^2 - y^2 = (y+4)(3y+4)$.

3.2. Splitting Middle Terms

3. $3(x+1) - x^2 - 2x - 1$
   We have
   $$\begin{aligned}
   3(x+1) - x^2 - 2x - 1 &= 3(x+1) - (x^2 + 2x + 1) \\
   &= 3(x+1) - (x+1)^2 \\
   &= (x+1)[3 - (x+1)] \\
   &= (x+1)(3 - x - 1) \\
   &= (x+1)(2 - x) \\
   &= -(x+1)(x-2).
   \end{aligned}$$
   Therefore, $3(x+1) - x^2 - 2x - 1 = -(x+1)(x-2)$.

4. $(2x-5)^2 - 4x^2 + 25$
   We have
   $$\begin{aligned}
   (2x-5)^2 - 4x^2 + 25 &= (2x-5)^2 - (4x^2 - 25) \\
   &= (2x-5)^2 - \left[(2x)^2 - 5^2\right] \\
   &= (2x-5)^2 - (2x-5)(2x+5) \\
   &= (2x-5)[(2x-5) - (2x+5)] \\
   &= (2x-5)(2x - 5 - 2x - 5) \\
   &= -10(2x-5).
   \end{aligned}$$
   Therefore, $(2x-5)^2 - 4x^2 + 25 = -10(2x-5)$.

5. $16(x+3)^2 - (x-1)^2$
   We have
   $$\begin{aligned}
   16(x+3)^2 - (x-1)^2 &= [4(x+3)]^2 - (x-1)^2 \\
   &= [4(x+3) - (x-1)][4(x+3) + (x-1)] \\
   &= (4x + 12 - x + 1)(4x + 12 + x - 1) \\
   &= (3x + 13)(5x + 11).
   \end{aligned}$$
   Therefore, $16(x+3)^2 - (x-1)^2 = (3x+13)(5x+11)$.

6. $x^3 - x$
   We have
   $$x^3 - x = x(x^2 - 1) = x(x-1)(x+1).$$
   Therefore, $x^3 - x = x(x-1)(x+1)$.

7. $t^4 - 1$

 We have
$$t^4 - 1 = (t^2)^2 - 1^2$$
$$= (t^2 - 1)(t^2 + 1)$$
$$= (t-1)(t+1)(t^2+1).$$

 Therefore, $t^4 - 1 = (t-1)(t+1)(t^2+1)$.

8. $4x^2 - 16x$

 We have
$$4x^2 - 16x = 4x(x-4).$$

 Therefore, $4x^2 - 16x = 4x(x-4)$.

9. $(x+1)^2 - 2(x^2-1) + 2x + 2$

 We have
$$(x+1)^2 - 2(x^2-1) + 2x + 2$$
$$= (x+1)^2 - 2(x-1)(x+1) + 2(x+1)$$
$$= (x+1)[(x+1) - 2(x-1) + 2]$$
$$= (x+1)(x+1-2x+2+2)$$
$$= (x+1)(-x+5)$$
$$= -(x+1)(x-5).$$

 Therefore, $(x+1)^2 - 2(x^2-1) + 2x + 2 = -(x+1)(x-5)$.

10. $x^5 - x$

 We have
$$x^5 - x = x(x^4 - 1)$$
$$= x\left[(x^2)^2 - 1^2\right]$$
$$= x(x^2-1)(x^2+1)$$
$$= x(x-1)(x+1)(x^2+1).$$

 Therefore, $(x+1)^2 - (x^2-1) + 2x + 2 = x(x-1)(x+1)(x^2+1)$.

11. $x(y-1)^2 - 4x$

 We have
$$x(y-1)^2 - 4x = x\left[(y-1)^2 - 4\right]$$

3.2. Splitting Middle Terms

$$= x\left[(y-1)^2 - 2^2\right]$$
$$= x\left[(y-1) - 2\right]\left[(y-1) + 2\right]$$
$$= x(y - 1 - 2)(y - 1 + 2)$$
$$= x(y-3)(y+1).$$

Therefore, $x(y-1)^2 - 4x = x(y-3)(y+1)$.

Problem 8. Find the value of a such that

$$3x^3 + 2x^2 + 1 = (x+1)\left(3x^2 - ax + 1\right).$$

Solution. Find the value of a.
We have

$$3x^3 + 2x^2 + 1 = (x+1)\left(3x^2 - ax + 1\right)$$
$$= 3x^3 - ax^2 + x + 3x^2 - ax + 1$$
$$= 3x^3 + (3-a)x^2 + (1-a)x + 1.$$

Comparing the coefficients of x^2 and x in both sides of the above equation, we obtain

$$\begin{cases} 3 - a = 2 \\ 1 - a = 0 \end{cases}.$$

It follows that

$$a = 1.$$

Therefore, $a = 1$.

Problem 9. Factor the following expressions:

1. $x^2 + 3x + 2$;
2. $x^2 - 3x + 2$;
3. $x^2 - 5x + 4$;
4. $x^2 - 11x + 28$;
5. $x^2 - 7x + 10$;
6. $x^2 - 11x + 18$;
7. $x^2 - 7x + 12$;
8. $x^2 - 12x + 20$;
9. $x^2 - 8x + 16$;
10. $x^2 - 11x + 30$;
11. $2x^2 - 12x + 10$;
12. $2x^2 - 3x + 1$;
13. $4x^2 - 5x + 1$;
14. $8x^2 - x - 7$;

15. $6x^2 - 2x - 4$;

16. $7x^2 - 4x - 3$;

17. $3x^2 - 5x + 2$;

18. $6x^2 - x - 5$;

19. $10x^2 - x - 9$;

20. $8x^2 - 4x - 4$.

Solution. Factor the following expressions:

1. $x^2 + 3x + 2$
   We have
   $$x^2 + 3x + 2 = x^2 + x + 2x + 2$$
   $$= x(x+1) + 2(x+1)$$
   $$= (x+1)(x+2).$$

   Therefore, $x^2 + 3x + 2 = (x+1)(x+2)$.

2. $x^2 - 3x + 2$
   We have
   $$x^2 - 3x + 2 = x^2 - x - 2x + 2$$
   $$= x(x-1) - 2(x-1)$$
   $$= (x-1)(x-2).$$

   Therefore, $x^2 - 3x + 2 = (x-1)(x-2)$.

3. $x^2 - 5x + 4$
   We have
   $$x^2 - 5x + 4 = x^2 - x - 4x + 4$$
   $$= x(x-1) - 4(x-1)$$
   $$= (x-1)(x-4).$$

   Therefore, $x^2 - 5x + 4 = (x-1)(x-4)$.

4. $x^2 - 11x + 28$
   We have
   $$x^2 - 11x + 28 = x^2 - 4x - 7x + 28$$
   $$= x(x-4) - 7(x-4)$$
   $$= (x-4)(x-7).$$

   Therefore, $x^2 - 11x + 28 = (x-4)(x-7)$.

3.2. Splitting Middle Terms

5. $x^2 - 7x + 10$
   We have
   $$\begin{aligned}x^2 - 7x + 10 &= x^2 - 2x - 5x + 10 \\ &= x(x-2) - 5(x-2) \\ &= (x-2)(x-5).\end{aligned}$$
   Therefore, $x^2 - 7x + 10 = (x-2)(x-5)$.

6. $x^2 - 11x + 18$
   We have
   $$\begin{aligned}x^2 - 11x + 18 &= x^2 - 2x - 9x + 18 \\ &= x(x-2) - 9(x-2) \\ &= (x-2)(x-9).\end{aligned}$$
   Therefore, $x^2 - 11x + 18 = (x-2)(x-9)$.

7. $x^2 - 7x + 12$
   We have
   $$\begin{aligned}x^2 - 7x + 12 &= x^2 - 3x - 4x + 12 \\ &= x(x-3) - 4(x-3) \\ &= (x-3)(x-4).\end{aligned}$$
   Therefore, $x^2 - 7x + 12 = (x-3)(x-4)$.

8. $x^2 - 12x + 20$
   We have
   $$\begin{aligned}x^2 - 12x + 20 &= x^2 - 2x - 10x + 20 \\ &= x(x-2) - 10(x-2) \\ &= (x-2)(x-10).\end{aligned}$$
   Therefore, $x^2 - 12x + 20 = (x-2)(x-10)$.

9. $x^2 - 8x + 16$
   We have
   $$\begin{aligned}x^2 - 8x + 16 &= x^2 - 4x - 4x + 16 \\ &= x(x-4) - 4(x-4)\end{aligned}$$

$$= (x-4)(x-4)$$
$$= (x-4)^2.$$

Therefore, $x^2 - 8x + 16 = (x-4)^2$.

10. $x^2 - 11x + 30$
   We have
   $$x^2 - 11x + 30 = x^2 - 5x - 6x + 30$$
   $$= x(x-5) - 6(x-5)$$
   $$= (x-5)(x-6).$$

   Therefore, $x^2 - 11x + 30 = (x-5)(x-6)$.

11. $2x^2 - 12x + 10$
   We have
   $$2x^2 - 12x + 10 = 2(x^2 - 6x + 5)$$
   $$= 2(x^2 - x - 5x + 5)$$
   $$= 2[x(x-1) - 5(x-1)]$$
   $$= 2(x-1)(x-5).$$

   Therefore, $2x^2 - 12x + 10 = 2(x-1)(x-5)$.

12. $2x^2 - 3x + 1$
   We have
   $$2x^2 - 3x + 1 = 2x^2 - 2x - x + 1$$
   $$= 2x(x-1) - (x-1)$$
   $$= (x-1)(2x-1).$$

   Therefore, $2x^2 - 3x + 1 = (x-1)(2x-1)$.

13. $4x^2 - 5x + 1$
   We have
   $$4x^2 - 5x + 1 = 4x^2 - 4x - x + 1$$
   $$= 4x(x-1) - (x-1)$$
   $$= (x-1)(4x-1).$$

   Therefore, $4x^2 - 5x + 1 = (x-1)(4x-1)$.

3.2. Splitting Middle Terms

14. $8x^2 - x - 7$
We have
$$8x^2 - x - 7 = 8x^2 - 8x + 7x - 7$$
$$= 8x(x-1) + 7(x-1)$$
$$= (x-1)(8x+7).$$

Therefore, $8x^2 - x - 7 = (x-1)(8x+7)$.

15. $6x^2 - 2x - 4$
We have
$$6x^2 - 2x - 4 = 6x^2 - 6x + 4x - 4$$
$$= 6x(x-1) + 4(x-1)$$
$$= (x-1)(6x+4)$$
$$= 2(x-1)(3x+2).$$

Therefore, $6x^2 - 2x - 4 = 2(x-1)(3x+2)$.

16. $7x^2 - 4x - 3$
We have
$$7x^2 - 4x - 3 = 7x^2 - 7x + 3x - 3$$
$$= 7x(x-1) + 3(x-1)$$
$$= (x-1)(7x+3).$$

Therefore, $7x^2 - 4x - 3 = (x-1)(7x+3)$.

17. $3x^2 - 5x + 2$
We have
$$3x^2 - 5x + 2 = 3x^2 - 3x - 2x + 2$$
$$= 3x(x-1) - 2(x-1)$$
$$= (x-1)(3x-2).$$

Therefore, $3x^2 - 5x + 2 = (x-1)(3x-2)$.

18. $6x^2 - x - 5$
We have
$$6x^2 - x - 5 = 6x^2 - 6x + 5x - 5$$

$$= 6x(x-1) + 5(x-1)$$
$$= (x-1)(6x+5).$$

Therefore, $6x^2 - x - 5 = (x-1)(6x+5)$.

19. $10x^2 - x - 9$

   We have

   $$10x^2 - x - 9 = 10x^2 - 10x + 9x - 9$$
   $$= 10x(x-1) + 9(x-1)$$
   $$= (x-1)(10x+9).$$

   Therefore, $10x^2 - x - 9 = (x-1)(10x+9)$.

20. $8x^2 - 4x - 4$

   We have

   $$8x^2 - 4x - 4 = 8x^2 - 8x + 4x - 4$$
   $$= 8x(x-1) + 4(x-1)$$
   $$= (x-1)(8x+4)$$
   $$= 4(x-1)(2x+1).$$

   Therefore, $8x^2 - 4x - 4 = 4(x-1)(2x+1)$.

Problem 10. Calculate the following expressions:

1. $A = 4x - \dfrac{x-1}{x-2}$;

2. $B = x - 1 + \dfrac{3x}{x-1}$;

3. $C = 3x - 2 + \dfrac{2}{x-2}$;

4. $D = 4x - 1 - \dfrac{x-1}{2x-1}$;

5. $E = 3x - 5 - \dfrac{3x-1}{x-1}$;

6. $F = 2x - 2 + \dfrac{3x-1}{x+2}$;

3.2. Splitting Middle Terms

7. $G = -3x + 1 - \dfrac{1}{x-1}$;

8. $H = 4x - 2 - \dfrac{3x-1}{x-2}$;

9. $I = 4x - 3 - \dfrac{2x-1}{x-2}$;

10. $J = -7x + 1 + \dfrac{7x-1}{3x-1}$.

Solution. Calculate the following expressions:

1. $A = 4x - \dfrac{x-1}{x-2}$

We have
$$A = 4x - \frac{x-1}{x-2}$$
$$= \frac{4x(x-2) - (x-1)}{x-2}$$
$$= \frac{4x^2 - 8x - x + 1}{x-2}$$
$$= \frac{4x^2 - 9x + 1}{x-2}.$$

Therefore, $A = \dfrac{4x^2 - 9x + 1}{x-2}$.

2. $B = x - 1 + \dfrac{3x}{x-1}$

We have
$$B = x - 1 + \frac{3x}{x-1}$$
$$= \frac{(x-1)(x-1) + 3x}{x-1}$$
$$= \frac{x^2 - x - x + 1 + 3x}{x-1}$$
$$= \frac{x^2 + x + 1}{x-1}.$$

Therefore, $B = \dfrac{x^2 + x + 1}{x-1}$.

3. $C = 3x - 2 + \dfrac{2}{x-2}$

We have
$$C = 3x - 2 + \dfrac{2}{x-2}$$
$$= \dfrac{(3x-2)(x-2)+2}{x-2}$$
$$= \dfrac{3x^2 - 6x - 2x + 4 + 2}{x-2}$$
$$= \dfrac{3x^2 - 8x + 6}{x-2}.$$

Therefore, $C = \dfrac{3x^2 - 8x + 6}{x-2}$.

4. $D = 4x - 1 - \dfrac{x-1}{2x-1}$

We have
$$D = 4x - 1 - \dfrac{x-1}{2x-1}$$
$$= \dfrac{(4x-1)(2x-1)-(x-1)}{2x-1}$$
$$= \dfrac{8x^2 - 4x - 2x + 1 - x + 1}{2x-1}$$
$$= \dfrac{8x^2 - 7x + 2}{2x-1}.$$

Therefore, $D = \dfrac{8x^2 - 7x + 2}{2x-1}$.

5. $E = 3x - 5 - \dfrac{3x-1}{x-1}$

We have
$$E = 3x - 5 - \dfrac{3x-1}{x-1}$$
$$= \dfrac{(3x-5)(x-1)-(3x-1)}{x-1}$$
$$= \dfrac{3x^2 - 3x - 5x + 5 - 3x + 1}{x-1}$$
$$= \dfrac{3x^2 - 11x + 6}{x-1}.$$

3.2. Splitting Middle Terms

Therefore, $E = \dfrac{3x^2 - 11x + 6}{x - 1}$.

6. $F = 2x - 2 + \dfrac{3x - 1}{x + 2}$

   We have

   $$\begin{aligned} F &= 2x - 2 + \dfrac{3x - 1}{x + 2} \\ &= \dfrac{(2x - 2)(x + 2) + 3x - 1}{x + 2} \\ &= \dfrac{2x^2 + 4x - 2x - 4 + 3x - 1}{x + 2} \\ &= \dfrac{2x^2 + 5x - 5}{x + 2}. \end{aligned}$$

   Therefore, $F = \dfrac{2x^2 + 5x - 5}{x + 2}$.

7. $G = -3x + 1 - \dfrac{1}{x - 1}$

   We have

   $$\begin{aligned} G &= -3x + 1 - \dfrac{1}{x - 1} \\ &= \dfrac{(-3x + 1)(x - 1) - 1}{x - 1} \\ &= \dfrac{-3x^2 + 3x + x - 1 - 1}{x - 1} \\ &= \dfrac{-3x^2 + 4x - 2}{x - 1}. \end{aligned}$$

   Therefore, $G = -3x + 1 - \dfrac{1}{x - 1}$.

8. $H = 4x - 2 - \dfrac{3x - 1}{x - 2}$

   We have

   $$\begin{aligned} H &= 4x - 2 - \dfrac{3x - 1}{x - 2} \\ &= \dfrac{(4x - 2)(x - 2) - (3x - 1)}{x - 2} \end{aligned}$$

$$= \frac{4x^2 - 8x - 2x + 4 - 3x + 1}{x - 2}$$
$$= \frac{4x^2 - 13x + 5}{x - 2}.$$

Therefore, $H = \dfrac{4x^2 - 13x + 5}{x - 2}.$

9. $I = 4x - 3 - \dfrac{2x - 1}{x - 2}$

We have
$$I = 4x - 3 - \frac{2x - 1}{x - 2}$$
$$= \frac{(4x - 3)(x - 2) - (2x - 1)}{x - 2}$$
$$= \frac{4x^2 - 8x - 3x + 6 - 2x + 1}{x - 2}$$
$$= \frac{4x^2 - 13x + 7}{x - 2}.$$

Therefore, $I = \dfrac{4x^2 - 13x + 7}{x - 2}.$

10. $J = -7x + 1 + \dfrac{7x - 1}{3x - 1}$

We have
$$J = -7x + 1 + \frac{7x - 1}{3x - 1}$$
$$= \frac{(-7x + 1)(3x - 1) + 7x - 1}{3x - 1}$$
$$= \frac{-21x^2 + 7x + 3x - 1 + 7x - 1}{3x - 1}$$
$$= \frac{-21x^2 + 17x - 2}{3x - 1}.$$

Therefore, $J = \dfrac{-21x^2 + 17x - 2}{3x - 1}.$

Problem 11. Find the value of a, b and c such that

1. $\dfrac{3x^2 - 2x + 1}{x - 1} = ax + b + \dfrac{c}{x - 1};$

3.2. Splitting Middle Terms

2. $\dfrac{5x^2 - x - 1}{x - 1} = ax + b + \dfrac{c}{x - 1}$;

3. $\dfrac{x^2 - 2x + 4}{x + 2} = ax + b + \dfrac{c}{x + 2}$;

4. $\dfrac{x^2 - 6x + 3}{x + 2} = ax + b + \dfrac{c}{x + 2}$.

Solution. Find the value of a, b and c such that

1. $\dfrac{3x^2 - 2x + 1}{x - 1} = ax + b + \dfrac{c}{x - 1}$

Multiply both sides of the above equation by $x - 1$, we obtain

$$3x^2 - 2x + 1 = (x - 1)(ax + b) + c.$$

Then

$$\begin{aligned}3x^2 - 2x + 1 &= (x - 1)(ax + b) + c \\ &= ax^2 + bx - ax - b + c \\ &= ax^2 + (b - a)x + (-b + c).\end{aligned}$$

Comparing the left-handed side and the right-handed side polynomials, we obtain

$$\begin{cases} a = 3 & (1) \\ b - a = -2 & (2) \\ -b + c = 1 & (3) \end{cases}.$$

Substitute $a = 3$ into (2), we obtain

$$b - 3 = -2.$$

It follows that

$$b = 1.$$

Substitute $b = 1$ into (3), we obtain

$$-1 + c = 1.$$

Then $c = 2$.
Therefore, $a = 3, b = 1$ and $c = 2$.

Chapter 3. Methods in Factorization

2. $\dfrac{5x^2 - x - 1}{x - 1} = ax + b + \dfrac{c}{x - 1}$

Multiply both sides of the above equation by $x-1$, we obtain

$$5x^2 - x - 1 = (x - 1)(ax + b) + c.$$

It follows that

$$5x^2 - x - 1 = ax^2 + bx - ax - b + c$$
$$= ax^2 + (b - a)x + (-b + c).$$

Comparing the left-handed side and the right-handed side polynomials, we obtain

$$\begin{cases} a = 5 & (1) \\ b - a = -1 & (2) \\ -b + c = -1 & (3) \end{cases}.$$

Substitute $a = 5$ into (2), we obtain

$$b - 5 = -1.$$

It follows that

$$b = 4.$$

Substitute $b = 4$ into (3), we obtain

$$-4 + c = -1.$$

Then $c = 3$.
Therefore, $a = 5, b = 4$ and $c = 3$.

3. $\dfrac{x^2 - 2x + 4}{x + 2} = ax + b + \dfrac{c}{x + 2}$

Multiply both sides of the above equation by $x+2$, we obtain

$$x^2 - 2x + 4 = (x + 2)(ax + b) + c.$$

It follows that

$$x^2 - 2x + 4 = ax^2 + bx + 2ax + 2b + c$$
$$= ax^2 + (b + 2a)x + (2b + c).$$

88

3.2. Splitting Middle Terms

Comparing the left-handed side and the right-handed side polynomials, we obtain
$$\begin{cases} a = 1 & (1) \\ b + 2a = -2 & (2) \\ 2b + c = 4 & (3) \end{cases}.$$

Substitute $a = 1$ into (2), we obtain
$$b + 2 = -2.$$

It follows that
$$b = -4.$$

Substitute $b = 4$ into (3), we obtain
$$-8 + c = 4.$$

Then $c = 12$.
Therefore, $a = 1, b = 4$ and $c = 12$.

4. $\dfrac{x^2 - 6x + 3}{x + 2} = ax + b + \dfrac{c}{x + 2}$

Multiply both sides of the above equation by $x+2$, we obtain
$$\begin{aligned} x^2 - 6x + 3 &= (x + 2)(ax + b) + c \\ &= ax^2 + bx + 2ax + 2b + c \\ &= ax^2 + (b + 2a)x + (2b + c). \end{aligned}$$

Comparing the left-handed side and the right-handed side polynomials, we obtain
$$\begin{cases} a = 1 & (1) \\ b + 2a = -6 & (2) \\ 2b + c = 3 & (3) \end{cases}.$$

Substitute $a = 1$ into (2), we obtain
$$b + 2 = -6.$$

It follows that
$$b = -8.$$

Substitute $b = -8$ into (3), we obtain
$$-16 + c = 3.$$
Then $c = 19$.
Therefore, $a = 1, b = -8$ and $c = 19$.

Problem 12. Compute the following expressions:

1. $\dfrac{4}{x-1} - \dfrac{2}{x^2-1}$;

2. $\dfrac{7x}{x+1} - \dfrac{3x}{x^2-1}$;

3. $\dfrac{7x-1}{x^2-x} - \dfrac{3x-1}{x-1}$;

4. $\dfrac{8x-1}{x^2-x} - \dfrac{4x-1}{x} + \dfrac{3}{x-1}$;

5. $\dfrac{6x-1}{x^3-x} - \dfrac{3x-1}{x} + \dfrac{7}{x-1}$;

6. $\dfrac{7x-3}{x^3-x} + \dfrac{4x-1}{x+1} - \dfrac{5x-1}{x-1}$;

7. $\dfrac{7}{x^4+x^2+1} - \dfrac{8x-1}{x^2-x+1}$;

8. $\dfrac{9x-1}{x^4+x^2+1} + \dfrac{8x-1}{x^2-x+1} + \dfrac{7x-1}{x^2+x+1}$.

Solution. Compute the following expressions:

1. $\dfrac{4}{x-1} - \dfrac{2}{x^2-1}$

We have
$$\dfrac{4}{x-1} - \dfrac{2}{x^2-1} = \dfrac{4}{x-1} - \dfrac{2}{(x-1)(x+1)}$$
$$= \dfrac{4(x+1) - 2}{(x-1)(x+1)}$$
$$= \dfrac{4x + 4 - 2}{x^2 - 1}$$
$$= \dfrac{4x + 2}{x^2 - 1}.$$

Therefore, $\dfrac{4}{x-1} - \dfrac{2}{x^2-1} = \dfrac{4x+2}{x^2-1}.$

3.2. Splitting Middle Terms

2. $\dfrac{7x}{x+1} - \dfrac{3x}{x^2-1}$

We have
$$\dfrac{7x}{x+1} - \dfrac{3x}{x^2-1} = \dfrac{7x}{x+1} - \dfrac{3x}{(x-1)(x+1)}$$
$$= \dfrac{7x(x-1) - 3x}{x+1}$$
$$= \dfrac{7x^2 - 7x - 3x}{x+1}$$
$$= \dfrac{7x^2 - 10x}{x+1}.$$

Therefore, $\dfrac{7x}{x+1} - \dfrac{3x}{x^2-1} = \dfrac{7x^2 - 10x}{x+1}.$

3. $\dfrac{7x-1}{x^2-x} - \dfrac{3x-1}{x-1}$

We have
$$\dfrac{7x-1}{x^2-x} - \dfrac{3x-1}{x-1} = \dfrac{7x-1}{x(x-1)} - \dfrac{3x-1}{x-1}$$
$$= \dfrac{7x - 1 - x(3x-1)}{x(x-1)}$$
$$= \dfrac{7x - 1 - 3x^2 + x}{x(x-1)}$$
$$= \dfrac{-3x^2 + 8x - 1}{x^2 - x}.$$

Therefore, $\dfrac{7x-1}{x^2-x} - \dfrac{3x-1}{x-1} = \dfrac{-3x^2 + 8x - 1}{x^2 - x}.$

4. $\dfrac{8x-1}{x^2-x} - \dfrac{4x-1}{x} + \dfrac{3}{x-1}$

We have
$$\dfrac{8x-1}{x^2-x} - \dfrac{4x-1}{x} + \dfrac{3}{x-1}$$
$$= \dfrac{8x-1}{x(x-1)} - \dfrac{4x-1}{x} + \dfrac{3}{x-1}$$
$$= \dfrac{8x - 1 - (x-1)(4x-1) + 3x}{x(x-1)}$$

$$= \frac{8x - 1 - \left(4x^2 - x - 4x + 1\right) + 3x}{x^2 - x}$$

$$= \frac{8x - 1 - \left(4x^2 - 5x + 1\right) + 3x}{x^2 - x}$$

$$= \frac{8x - 1 - 4x^2 + 5x - 1 + 3x}{x^2 - x}$$

$$= \frac{-4x^2 + 16x - 2}{x^2 - x}.$$

Therefore, $\dfrac{8x - 1}{x^2 - x} - \dfrac{4x - 1}{x} + \dfrac{3}{x - 1} = \dfrac{-4x^2 + 16x - 2}{x^2 - x}.$

5. $\dfrac{6x - 1}{x^3 - x} - \dfrac{3x - 1}{x} + \dfrac{7}{x - 1}$

$$\frac{6x - 1}{x^3 - x} - \frac{3x - 1}{x} + \frac{7}{x - 1}$$

$$= \frac{6x - 1}{x\left(x^2 - 1\right)} - \frac{3x - 1}{x} + \frac{7}{x - 1}$$

$$= \frac{6x - 1}{x\left(x - 1\right)\left(x + 1\right)} - \frac{3x - 1}{x} + \frac{7}{x - 1}$$

$$= \frac{6x - 1 - (x - 1)(x + 1)(3x - 1) + 7x(x + 1)}{x(x - 1)(x + 1)}$$

$$= \frac{6x - 1 - \left(x^2 - 1\right)(3x - 1) + 7x^2 + 7x}{x^3 - x}$$

$$= \frac{6x - 1 - \left(3x^3 - x^2 - 3x + 1\right) + 7x^2 + 7x}{x^3 - x}$$

$$= \frac{6x - 1 - 3x^3 + x^2 + 3x - 1 + 7x^2 + 7x}{x^3 - x}$$

$$= \frac{-3x^3 + 8x^2 + 16x - 2}{x^3 - x}.$$

Therefore, $\dfrac{6x - 1}{x^3 - x} - \dfrac{3x - 1}{x} + \dfrac{7}{x - 1} = \dfrac{-3x^3 + 8x^2 + 16x - 2}{x^3 - x}.$

6. $\dfrac{7x - 3}{x^3 - x} + \dfrac{4x - 1}{x + 1} - \dfrac{5x - 1}{x - 1}$
We have

$$\frac{7x - 3}{x^3 - x} + \frac{4x - 1}{x + 1} - \frac{5x - 1}{x - 1}$$

3.2. Splitting Middle Terms

$$= \frac{7x-3}{x(x^2-1)} + \frac{4x-1}{x+1} - \frac{5x-1}{x-1}$$

$$= \frac{7x-3}{x(x-1)(x+1)} + \frac{4x-1}{x+1} - \frac{5x-1}{x-1}$$

$$= \frac{7x-3+x(x-1)(4x-1)-x(x+1)(5x-1)}{x(x-1)(x+1)}$$

$$= \frac{7x-3+(x^2-x)(4x-1)-(x^2+x)(5x-1)}{x^3-x}$$

$$= \frac{7x-3+4x^3-x^2-4x^2+x-(5x^3-x^2+5x^2-x)}{x^3-x}$$

$$= \frac{4x^3-5x^2+8x-3-5x^3-4x^2+x}{x^3-x}$$

$$= \frac{-x^3-9x^2+9x-3}{x^3-x}.$$

Therefore, $\dfrac{7x-3}{x^3-x} + \dfrac{4x-1}{x+1} - \dfrac{5x-1}{x-1} = \dfrac{-x^3-9x^2+9x-3}{x^3-x}$.

7. $\dfrac{7}{x^4+x^2+1} - \dfrac{8x-1}{x^2-x+1}$

We have

$$x^4 + x^2 + 1 = x^4 + 2x^2 + 1 - x^2$$
$$= (x^2+1)^2 - x^2$$
$$= (x^2+1-x)(x^2+1+x)$$
$$= (x^2-x+1)(x^2+x+1).$$

It follows that

$$\frac{7}{x^4+x^2+1} - \frac{8x-1}{x^2-x+1}$$

$$= \frac{7}{(x^2-x+1)(x^2+x+1)} - \frac{8x-1}{x^2-x+1}$$

$$= \frac{7-(8x-1)(x^2+x+1)}{(x^2-x+1)(x^2+x+1)}$$

$$= \frac{7-(8x^3+8x^2+8x-x^2-x-1)}{x^4+x^2+1}$$

$$= \frac{7-8x^3-7x^2-7x+1}{x^4+x^2+1}$$

Chapter 3. Methods in Factorization

$$= \frac{-8x^3 - 7x^2 - 7x + 8}{x^4 + x^2 + 1}.$$

Therefore, $\dfrac{7}{x^4 + x^2 + 1} - \dfrac{8x - 1}{x^2 - x + 1} = \dfrac{-8x^3 - 7x^2 - 7x + 8}{x^4 + x^2 + 1}.$

8. $\dfrac{9x - 1}{x^4 + x^2 + 1} + \dfrac{8x - 1}{x^2 - x + 1} + \dfrac{7x - 1}{x^2 + x + 1}$

We have

$$\frac{9x - 1}{x^4 + x^2 + 1} + \frac{8x - 1}{x^2 - x + 1} + \frac{7x - 1}{x^2 + x + 1}$$
$$= \frac{9x - 1}{(x^2 - x + 1)(x^2 + x + 1)} + \frac{8x - 1}{x^2 - x + 1} + \frac{7x - 1}{x^2 + x + 1}$$
$$= \frac{9x - 1 + (8x - 1)(x^2 + x + 1) + (7x - 1)(x^2 - x + 1)}{(x^2 - x + 1)(x^2 + x + 1)}$$
$$= \frac{15x^3 - x^2 + 24x - 3}{x^4 + x^2 + 1}.$$

Therefore,

$$\frac{9x - 1}{x^4 + x^2 + 1} + \frac{8x - 1}{x^2 - x + 1} + \frac{7x - 1}{x^2 + x + 1}$$
$$= \frac{15x^3 - x^2 + 24x - 3}{x^4 + x^2 + 1}.$$

Problem 13. Compute the following expressions:

1. $A = \dfrac{\dfrac{1}{x} + \dfrac{1}{x-1}}{\dfrac{2x+1}{x+1} + \dfrac{3}{x-1}};$

2. $B = \dfrac{\dfrac{3x-1}{x^2-1} + \dfrac{4x-1}{x-1}}{\dfrac{3x-1}{x+1} - \dfrac{4}{x^2-1}};$

3. $C = \dfrac{\dfrac{1}{x} + \dfrac{1}{x-1} + \dfrac{1}{x^2-x}}{\dfrac{2x-1}{x^2-x} + \dfrac{3}{x}};$

4. $D = \dfrac{\dfrac{1}{x} + \dfrac{1}{x+1} + \dfrac{3}{x-1}}{\dfrac{4x-1}{x^2-x} + \dfrac{3x-1}{x^2-1}};$

5. $E = \dfrac{x + \dfrac{1}{x + \dfrac{1}{x + \dfrac{1}{x}}}}{x}.$

Solution. Compute the following expressions:

3.2. Splitting Middle Terms

1. $A = \dfrac{\dfrac{1}{x} + \dfrac{1}{x-1}}{\dfrac{2x+1}{x+1} + \dfrac{3}{x-1}}$

   We have
   $$A = \dfrac{\dfrac{1}{x} + \dfrac{1}{x-1}}{\dfrac{2x+1}{x+1} + \dfrac{3}{x-1}}$$
   $$= \dfrac{\dfrac{x-1+x}{x(x-1)}}{\dfrac{(2x+1)(x-1)+3(x+1)}{(x+1)(x-1)}}$$
   $$= \dfrac{2x-1}{x(x-1)} \times \dfrac{(x+1)(x-1)}{2x^2-2x+x-1+3x+3}$$
   $$= \dfrac{(2x-1)(x+1)}{x(2x^2+2x+2)}$$
   $$= \dfrac{(2x-1)(x+1)}{2x(x^2+x+1)}.$$

   Therefore, $A = \dfrac{(2x-1)(x+1)}{2x(x^2+x+1)}.$

2. $B = \dfrac{\dfrac{3x-1}{x^2-1} + \dfrac{4x-1}{x-1}}{\dfrac{3x-1}{x+1} - \dfrac{4}{x^2-1}}$

   We have
   $$B = \dfrac{\dfrac{3x-1}{x^2-1} + \dfrac{4x-1}{x-1}}{\dfrac{3x-1}{x+1} - \dfrac{4}{x^2-1}}$$
   $$= \dfrac{\dfrac{3x-1}{(x-1)(x+1)} + \dfrac{4x-1}{x-1}}{\dfrac{3x-1}{x+1} - \dfrac{4}{(x-1)(x+1)}}$$
   $$= \dfrac{\dfrac{3x-1+(4x-1)(x+1)}{(x-1)(x+1)}}{\dfrac{(3x-1)(x-1)-4}{(x-1)(x+1)}}$$

$$= \frac{3x - 1 + 4x^2 + 4x - x - 1}{(x-1)(x+1)} \times \frac{(x-1)(x+1)}{3x^2 - 3x - x + 1 - 4}$$

$$= \frac{4x^2 + 6x - 2}{3x^2 - 4x - 3}$$

$$= \frac{2\left(2x^2 + 3x - 1\right)}{3x^2 - 4x - 3}.$$

Therefore, $B = \dfrac{2\left(2x^2 + 3x - 1\right)}{3x^2 - 4x - 3}$.

3. $C = \dfrac{\dfrac{1}{x} + \dfrac{1}{x-1} + \dfrac{1}{x^2 - x}}{\dfrac{2x-1}{x^2 - x} + \dfrac{3}{x}}$

We have

$$C = \frac{\dfrac{1}{x} + \dfrac{1}{x-1} + \dfrac{1}{x^2 - x}}{\dfrac{2x-1}{x^2 - x} + \dfrac{3}{x}}$$

$$= \frac{\dfrac{1}{x} + \dfrac{1}{x-1} + \dfrac{1}{x(x-1)}}{\dfrac{2x-1}{x(x-1)} + \dfrac{3}{x}}$$

$$= \frac{\dfrac{x - 1 + x + 1}{x(x-1)}}{\dfrac{2x - 1 + 3(x-1)}{x(x-1)}}$$

$$= \frac{2x}{x(x-1)} \times \frac{x(x-1)}{2x - 1 + 3x - 3}$$

$$= \frac{2x}{5x - 4}.$$

Therefore, $C = \dfrac{2x}{5x - 4}$.

4. $D = \dfrac{\dfrac{1}{x} + \dfrac{1}{x+1} + \dfrac{3}{x-1}}{\dfrac{4x-1}{x^2 - x} + \dfrac{3x-1}{x^2 - 1}}$

3.2. Splitting Middle Terms

We have
$$D = \frac{\dfrac{1}{x} + \dfrac{1}{x+1} + \dfrac{3}{x-1}}{\dfrac{4x-1}{x^2-x} + \dfrac{3x-1}{x^2-1}}$$

$$= \frac{\dfrac{(x+1)(x-1) + x(x-1) + 3x(x+1)}{x(x+1)(x-1)}}{\dfrac{4x-1}{x(x-1)} + \dfrac{3x-1}{(x-1)(x+1)}}$$

$$= \frac{\dfrac{x^2 - 1 + x^2 - x + 3x^2 + 3x}{x(x+1)(x-1)}}{\dfrac{(4x-1)(x+1) + x(3x-1)}{x(x-1)(x+1)}}$$

$$= \frac{5x^2 + 2x - 1}{x(x+1)(x-1)} \times \frac{x(x-1)(x+1)}{4x^2 + 4x - x - 1 + 3x^2 - x}$$

$$= \frac{5x^2 + 2x - 1}{7x^2 + 2x - 1}$$

Therefore, $D = \dfrac{5x^2 + 2x - 1}{7x^2 + 2x - 1}$.

5. $E = \dfrac{x + \dfrac{1}{x + \dfrac{1}{x + \dfrac{1}{x}}}}{x}$

We have

$$E = \frac{x + \dfrac{1}{x + \dfrac{1}{x + \dfrac{1}{x}}}}{x}$$

$$= \frac{x + \dfrac{1}{x + \dfrac{1}{\frac{x^2+1}{x}}}}{x}$$

$$= \frac{x + \dfrac{1}{x + \dfrac{x}{x^2+1}}}{x}$$

$$= \frac{1}{\frac{x^3+x+x}{x^2+1}}$$
$$= \frac{x^2+1}{x^3+2x} \times \frac{1}{x}$$
$$= \frac{x^2+1}{x^4+2x^2}.$$

Therefore, $E = \dfrac{x^2+1}{x^4+2x^2}$.

Problem 14. Find the value of $A = \dfrac{2x^3-8x}{4x^3-16x^2+16x}$ for $x = 2024$.

Solution. Find the value of A.
We have
$$A = \frac{2x^3-8x}{4x^3-16x^2+16x}$$
$$= \frac{2x(x^2-4)}{4x(x^2-4x+4)}$$
$$= \frac{(x-2)(x+2)}{2(x-2)^2}$$
$$= \frac{x+2}{2(x-2)}.$$

Substitute x by 2024, we obtain
$$A = \frac{2024+2}{2(2024-2)}$$
$$= \frac{2026}{2(2022)}$$
$$= \frac{1013}{2022}.$$

Therefore, $A = \dfrac{1013}{2022}$.

Chapter 4

Basic Algebraic Identities For Factorization

Algebraic identities are foundational tools in algebra that simplify expressions and solve equations efficiently. They are universally true equations involving algebraic expressions that hold for all values of the variables involved. These identities provide a structured way to manipulate and transform algebraic expressions, making complex problems more manageable. In this chapter, we will introduce readers some important identities that are used to factor algebraic expressions.

4.1 Perfect Square or Perfect Cube Identities

We start this chapter by introduce readers to the most important identities for expanding power expressions.

> **Theorem 1**
> For all real numbers a and b, we obtain the following identities:
> 1. $(a+b)^2 = a^2 + 2ab + b^2$;

Chapter 4. Basic Algebraic Identities For Factorization

2. $(a-b)^2 = a^2 - 2ab + b^2$;
3. $(a+b)^3 = a^3 + 3a^2b + 3ab^2 + b^3$;
4. $(a-b)^3 = a^3 - 3a^2b + 3ab^2 - b^3$.

Notice that we use the above identities to expand the power of algebraic expressions.

Proof. 1. This identity is already proved in the previous chapter.

2. $(a-b)^2 = a^2 - 2ab + b^2$
We have
$$(a-b)^2 = [a + (-b)]^2$$
$$= a^2 + 2(a)(-b) + (-b)^2$$
$$= a^2 - 2ab + b^2.$$

Therefore, $(a-b)^2 = a^2 - 2ab + b^2$.

3. This identity is already proved in the previous chapter.

4. $(a-b)^3 = a^3 - 3a^2b + 3ab^2 - b^3$
Observe that
$$(a-b)^3 = [a + (-b)]^3$$
$$= a^3 + 3a^2(-b) + 3a(-b)^2 + (-b)^3$$
$$= a^3 - 3a^2b + 3ab^2 - b^3.$$

Therefore, $(a-b)^3 = a^3 - 3a^2b + 3ab^2 - b^3$. □

Example 30

Expand the following expressions:

4.1. Perfect Square or Perfect Cube Identities

1. $(x+1)^2$;
2. $(x+2)^2$;
3. $(x+3)^2$;
4. $(x+4)^2$;
5. $(x+5)^2$;
6. $(x-1)^2$;
7. $(x-2)^2$;
8. $(x-3)^2$;
9. $(x-4)^2$;
10. $(x-5)^2$;
11. $(x+1)^3$;
12. $(x+2)^3$;
13. $(x+3)^3$;
14. $(x+4)^3$;
15. $(x+5)^3$;
16. $(x-1)^3$;
17. $(x-2)^3$;
18. $(x-3)^3$;
19. $(x-4)^3$;
20. $(x-5)^3$.

Solution. Expand the expressions:

1. $(x+1)^2$
We have
$$(x+1)^2 = x^2 + 2(x)(1) + 1^2$$
$$= x^2 + 2x + 1.$$

2. $(x+2)^2$
We have
$$(x+2)^2 = x^2 + 2(x)(2) + 2^2$$
$$= x^2 + 4x + 4.$$

3. $(x+3)^2$
We have
$$(x+3)^2 = x^2 + 2(x)(3) + 3^2$$
$$= x^2 + 6x + 9.$$

4. $(x+4)^2$
We have
$$(x+4)^2 = x^2 + 2(x)(4) + 4^2$$

$$= x^2 + 8x + 16.$$

5. $(x+5)^2$
   We have
   $$(x+5)^2 = x^2 + 2(x)(5) + 5^2$$
   $$= x^2 + 10x + 25.$$

6. $(x-1)^2$
   We have
   $$(x-1)^2 = x^2 - 2(x)(1) + 1^2$$
   $$= x^2 - 2x + 1.$$

7. $(x-2)^2$
   We have
   $$(x-2)^2 = x^2 - 2(x)(2) + 2^2$$
   $$= x^2 - 4x + 4.$$

8. $(x-3)^2$
   We have
   $$(x-3)^2 = x^2 - 2(x)(3) + 3^2$$
   $$= x^2 - 6x + 9.$$

9. $(x-4)^2$
   We have
   $$(x-4)^2 = x^2 - 2(x)(4) + 4^2$$
   $$= x^2 - 8x + 16.$$

10. $(x-5)^2$
    We have
    $$(x-5)^2 = x^2 - 2(x)(5) + 5^2$$
    $$= x^2 - 10x + 25.$$

4.1. Perfect Square or Perfect Cube Identities

11. $(x+1)^3$
    We have
    $$(x+1)^3 = x^3 + 3(x^2)(1) + 3(x)(1^2) + 1^3$$
    $$= x^3 + 3x^2 + 3x + 1.$$

12. $(x+2)^3$
    We have
    $$(x+2)^3 = x^3 + 3(x^2)(2) + 3(x)(2^2) + 2^3$$
    $$= x^3 + 6x^2 + 12x + 8.$$

13. $(x+3)^3$
    We have
    $$(x+3)^3 = x^3 + 3(x^2)(3) + 3(x)(3^2) + 3^3$$
    $$= x^3 + 9x^2 + 27x + 27.$$

14. $(x+4)^3$
    We have
    $$(x+4)^3 = x^3 + 3(x^2)(4) + 3(x)(4^2) + 4^3$$
    $$= x^3 + 12x^2 + 48x + 64.$$

15. $(x+5)^3$
    We have
    $$(x+5)^3 = x^3 + 3(x^2)(5) + 3(x)(5^2) + 5^3$$
    $$= x^3 + 15x^2 + 75x + 125.$$

16. $(x+1)^3$
    We have
    $$(x-1)^3 = x^3 - 3(x^2)(1) + 3(x)(1^2) - 1^3$$
    $$= x^3 - 3x^2 + 3x - 1.$$

17. $(x-2)^3$
    We have
    $$(x-2)^3 = x^3 - 3(x^2)(2) + 3(x)(2^2) - 2^3$$
    $$= x^3 - 6x^2 + 12x - 8.$$

Chapter 4. Basic Algebraic Identities For Factorization

18. $(x-3)^3$

    We have
    $$(x-3)^3 = x^3 - 3(x^2)(3) + 3(x)(3^2) - 3^3$$
    $$= x^3 - 9x^2 + 27x - 27.$$

19. $(x-4)^3$

    We have
    $$(x-4)^3 = x^3 - 3(x^2)(4) + 3(x)(4^2) - 4^3$$
    $$= x^3 - 12x^2 + 48x - 64.$$

20. $(x-5)^3$

    We have
    $$(x-5)^3 = x^3 - 3(x^2)(5) + 3(x)(5^2) - 5^3$$
    $$= x^3 - 15x^2 + 75x - 125.$$

Practice 1

Expand the following expressions:

1. $(2x+1)^2$;
2. $(2x+3)^2$;
3. $(3x+2)^2$;
4. $(4x+1)^2$;
5. $(5x+3)^2$;
6. $(2x-5)^2$;
7. $(4x-1)^2$;
8. $(2x-7)^2$;
9. $(6x-2)^2$;
10. $(7x-3)^2$;
11. $(3x-1)^2$;
12. $(2x+1)^3$;
13. $(2x+3)^3$;
14. $(3x+2)^3$;
15. $(4x+1)^3$;
16. $(5x+3)^3$;
17. $(2x-5)^3$;
18. $(4x-1)^3$;
19. $(2x-7)^3$;
20. $(6x-2)^3$;
21. $(7x-3)^3$;
22. $(3x-1)^3$.

4.1. Perfect Square or Perfect Cube Identities

4.1.1 Factoring Identities

> **Theorem 2**
> For all real numbers a and b, we obtain the following factoring identities:
> 1. $a^2 - b^2 = (a-b)(a+b)$;
> 2. $a^3 - b^3 = (a-b)(a^2 + ab + b^2)$;
> 3. $a^3 + b^3 = (a+b)(a^2 - ab + b^2)$.

Proof. 1. $a^2 - b^2 = (a-b)(a+b)$
Observe that
$$(a-b)(a+b) = a^2 + ab - ab - b^2$$
$$= a^2 - b^2.$$

2. $a^3 - b^3 = (a-b)(a^2 + ab + b^2)$
We have
$$(a-b)(a^2 + ab + b^2) = a^3 + a^2b + ab^2 - a^2b - ab^2 - b^3$$
$$= a^3 - b^3.$$

3. $a^3 + b^3 = (a+b)(a^2 - ab + b^2)$
Observe that
$$(a+b)(a^2 - ab + b^2) = a^3 - a^2b + ab^2 + a^2b - ab^2 + b^3$$
$$= a^3 + b^3.$$

$\square$

The readers should memorize the above identities. They are the powerful tools in factorization. The provided examples below will help the readers about how to use the Factoring Identities.

> **Example 31**
> Factor the following expressions:

Chapter 4. Basic Algebraic Identities For Factorization

1. $x^2 - 1$;
2. $x^2 - 9^2$;
3. $4x^2 - a^2$;
4. $a^2 - (b+c)^2$;
5. $(a+b)^2 - (c+d)^2$;
6. $x^3 - 1$;
7. $x^3 - 8$;
8. $8x^3 - 27$;
9. $27x^3 - 8$;
10. $a^3x^3 - b^3$;
11. $x^3 + 1$;
12. $x^3 + 8$;
13. $8x^3 + 27$;
14. $27x^3 + 8$;
15. $a^3x^3 + b^3$.

Solution. 1. $x^2 - 1$
We have $x^2 - 1 = x^2 - 1^2 = (x-1)(x+1)$.

2. $x^2 - 9$
We have $x^2 - 9 = x^2 - 3^2 = (x-3)(x+3)$.

3. $4x^2 - a^2$
We have $4x^2 - a^2 = 2^2 x^2 - a^2 = (2x)^2 - a^2 = (2x - a)(2x + a)$.

4. $a^2 - (b+c)^2$
We have
$$a^2 - (b+c)^2 = [a - (b+c)][a + (b+c)]$$
$$= (a - b - c)(a + b + c).$$

5. $(a+b)^2 - (c+d)^2$
We have
$$(a+b)^2 - (c+d)^2 = [(a+b) - (c+d)][(a+b) + (c+d)]$$
$$= (a + b - c - d)(a + b + c + d).$$

6. $x^3 - 1$
We have
$$x^3 - 1 = x^3 - 1^3$$
$$= (x-1)\left[x^2 + (x)(1) + 1^2\right]$$
$$= (x-1)\left(x^2 + x + 1\right).$$

4.1. Perfect Square or Perfect Cube Identities

7. $x^3 - 8$
   We have
   $$\begin{aligned} x^3 - 8 &= x^3 - 2^3 \\ &= (x-2)\left[x^2 + (x)(2) + 2^2\right] \\ &= (x-2)\left(x^2 + 2x + 4\right). \end{aligned}$$

8. $8x^3 - 27$
   We have
   $$\begin{aligned} 8x^3 - 27 &= 2^3 x^3 - 3^3 \\ &= (2x)^3 - 3^3 \\ &= (2x-3)\left[(2x)^2 + (2x)(3) + 3^2\right] \\ &= (2x-3)\left(4x^2 + 6x + 9\right). \end{aligned}$$

9. $27x^3 - 8$
   We have
   $$\begin{aligned} 27x^3 - 8 &= 3^3 x^3 - 2^3 \\ &= (3x)^3 - 2^3 \\ &= (3x-2)\left[(3x)^2 + (3x)(2) + 2^2\right] \\ &= (3x-2)\left(9x^2 + 6x + 4\right). \end{aligned}$$

10. $a^3 x^3 - b^3$
    We have
    $$\begin{aligned} a^3 x^3 - b^3 &= (ax)^3 - b^3 \\ &= (ax-b)\left[(ax)^2 + (ax)b + b^2\right] \\ &= (ax-b)\left(a^2 x^2 + abx + b^2\right). \end{aligned}$$

11. $x^3 + 1$
    We have
    $$\begin{aligned} x^3 + 1 &= x^3 + 1^3 \\ &= (x+1)\left[x^2 - (x)(1) + 1^2\right] \\ &= (x+1)\left(x^2 - x + 1\right). \end{aligned}$$

Chapter 4. Basic Algebraic Identities For Factorization

12. $x^3 + 8$

    We have

    $$\begin{aligned} x^3 + 8 &= x^3 + 2^3 \\ &= (x+2)\left[x^2 - (x)(2) + 2^2\right] \\ &= (x+2)\left(x^2 - 2x + 4\right). \end{aligned}$$

13. $8x^3 + 27$

    We have

    $$\begin{aligned} 8x^3 + 27 &= 2^3 x^3 + 3^3 \\ &= (2x)^3 + 3^3 \\ &= (2x+3)\left[(2x)^2 - (2x)(3) + 3^2\right] \\ &= (2x+3)\left(4x^2 - 6x + 9\right). \end{aligned}$$

14. $27x^3 + 8$

    We have

    $$\begin{aligned} 27x^3 + 8 &= 3^3 x^3 + 2^3 \\ &= (3x)^3 + 2^3 \\ &= (3x+2)\left[(3x)^2 - (3x)(2) + 2^2\right] \\ &= (3x+2)\left(9x^2 - 6x + 4\right). \end{aligned}$$

15. $a^3 x^3 + b^3$

    We have

    $$\begin{aligned} a^3 x^3 + b^3 &= (ax)^3 + b^3 \\ &= (ax+b)\left[(ax)^2 - (ax)b + b^2\right] \\ &= (ax+b)\left(a^2 x^2 - abx + b^2\right). \end{aligned}$$

> **Practice 2**
>
> Factor the following expressions:

4.1. Perfect Square or Perfect Cube Identities

1. $a^2 + 2ab + b^2 - c^2$;
2. $x^2 + 2x + 1 - y^2$;
3. $(2x+1)^2 - (3x-2)^2$;
4. $(a+b)^2 - (2a-b)^2$;
5. $(x+1)^3 - 8$;
6. $(2x)^3 - 125$;
7. $2x^3 - 16$;
8. $8x^3 - 64$;
9. $x^3 + 3x^2 + 3x - 7$;
10. $(x+1)^3 + 8$;
11. $(2x)^3 + 125$;
12. $2x^3 + 16$;
13. $8x^3 + 64$;
14. $x^3 + 3x^2 + 3x + 9$.

Chapter 4. Basic Algebraic Identities For Factorization

Exercises

Problem 1. Simplify
$$P = (2+1)\left(2^2+1\right)\left(2^4+1\right)\left(2^8+1\right)\left(2^{16}+1\right)\left(2^{32}+1\right).$$

Problem 2. Factor the following expressions:
$$a^2(b-c) + b^2(c-a) + c^2(a-b).$$

Problem 3. Suppose that x, y and z are three real numbers that satisfy
$$\begin{cases} x^2 + 2y + 1 = 0 \\ y^2 + 2z + 1 = 0 \\ z^2 + 2x + 1 = 0 \end{cases}.$$
Find the value of $A = x^n + y^n + z^n$, where n is a positive integer.

Problem 4. For all real numbers x, y and z, prove the following identities:

1. $(x+y)(y+z)(z+x) = (x+y+z)(xy+yz+zx) - xyz$;
2. $(x+y+z)^2 = x^2 + y^2 + z^2 + 2(xy+yz+zx)$;
3. $x^3 + y^3 + z^3 - 3xyz = (x+y+z)(x^2+y^2+z^2-xy-yz-zx)$.

Problem 5. Given three real numbers a, b and c such that $\frac{1}{a} + \frac{1}{b} + \frac{1}{c} = 0$ and $a+b+c = 1$. Compute $a^2 + b^2 + c^2$.

Problem 6. Let a, b and c be three distinct rational numbers. Prove that
$$\frac{1}{(a-b)^2} + \frac{1}{(b-c)^2} + \frac{1}{(c-a)^2}$$
is a square of a rational number.

Problem 7. Let a, b and c be three real number such that $a+b+c = 0$. Prove that

$$\left(\frac{a-b}{c} + \frac{b-c}{a} + \frac{c-a}{b}\right)\left(\frac{c}{a-b} + \frac{a}{b-c} + \frac{b}{c-a}\right) = 9.$$

Chapter 5

Solutions

Problem 1. Simplify
$$P = (2+1)\left(2^2+1\right)\left(2^4+1\right)\left(2^8+1\right)\left(2^{16}+1\right)\left(2^{32}+1\right).$$

Solution. Simplify P.
We have

$$\begin{aligned}
P &= (2+1)\left(2^2+1\right)\left(2^4+1\right)\left(2^8+1\right)\left(2^{16}+1\right)\left(2^{32}+1\right) \\
&= (2-1)(2+1)\left(2^2+1\right)\left(2^4+1\right)\left(2^8+1\right)\left(2^{16}+1\right)\left(2^{32}+1\right) \\
&= \left(2^2-1\right)\left(2^2+1\right)\left(2^4+1\right)\left(2^8+1\right)\left(2^{16}+1\right)\left(2^{32}+1\right) \\
&= \left(2^4-1\right)\left(2^4+1\right)\left(2^8+1\right)\left(2^{16}+1\right)\left(2^{32}+1\right) \\
&= \left(2^8-1\right)\left(2^8+1\right)\left(2^{16}+1\right)\left(2^{32}+1\right) \\
&= \left(2^{16}-1\right)\left(2^{16}+1\right)\left(2^{32}+1\right) \\
&= \left(2^{32}-1\right)\left(2^{32}+1\right) \\
&= 2^{64}-1.
\end{aligned}$$

Problem 2. Factor the following expressions:
$$a^2(b-c) + b^2(c-a) + c^2(a-b).$$

Solution. Factor the following expressions:
We have

$$\begin{aligned}
&a^2(b-c) + b^2(c-a) + c^2(a-b) \\
&= a^2 b - a^2 c + b^2 c - ab^2 + c^2(a-b)
\end{aligned}$$

$$\begin{aligned}
&= (a^2b - ab^2) + (-a^2c + b^2c) + c^2(a-b) \\
&= ab(a-b) - c(a^2 - b^2) + c^2(a-b) \\
&= ab(a-b) - c(a-b)(a+b) + c^2(a-b) \\
&= (a-b)\left[ab - c(a+b) + c^2\right] \\
&= (a-b)(ab - ac - bc + c^2) \\
&= (a-b)[a(b-c) - c(b-c)] \\
&= (a-b)(b-c)(a-c).
\end{aligned}$$

Problem 3. Suppose that x, y and z are three real numbers that satisfy
$$\begin{cases} x^2 + 2y + 1 = 0 \\ y^2 + 2z + 1 = 0 \\ z^2 + 2x + 1 = 0 \end{cases}.$$
Find the value of $A = x^n + y^n + z^n$, where n is a positive integer.

Solution. Find the value of $A = x^n + y^n + z^n$.
We have
$$\begin{cases} x^2 + 2y + 1 = 0 \\ y^2 + 2z + 1 = 0 \\ z^2 + 2x + 1 = 0 \end{cases}.$$
Adding the equalities, we obtain
$$(x^2 + 2x + 1) + (y^2 + 2y + 1) + (z^2 + 2z + 1) = 0$$
$$(x+1)^2 + (y+1)^2 + (z+1)^2 = 0.$$
It follows that
$$\begin{cases} x+1 = 0 \\ y+1 = 0 \\ z+1 = 0 \end{cases} \text{ or } x = y = z = -1.$$
It implies that $A = (-1)^n + (-1)^n + (-1)^n$.

- If n is an even number, we obtain $A = 1 + 1 + 1 = 3$.
- If n is an odd number, we obtain $A = -1 - 1 - 1 = -3$.

Problem 4. For all real numbers x, y and z, prove the following identities:

1. $(x+y)(y+z)(z+x) = (x+y+z)(xy+yz+zx) - xyz$;
2. $(x+y+z)^2 = x^2 + y^2 + z^2 + 2(xy+yz+zx)$;
3. $x^3 + y^3 + z^3 - 3xyz = (x+y+z)(x^2+y^2+z^2 - xy - yz - zx)$.

Solution. Prove the identities:

1. $(x+y)(y+z)(z+x) = (x+y+z)(xy+yz+zx) - xyz$
   We have

$$(x+y)(y+z)(z+x)$$
$$= (xy + xz + y^2 + yz)(z+x)$$
$$= xyz + x^2y + xz^2 + x^2z + y^2z + xy^2 + yz^2 + xyz$$
$$= (x^2y + xy^2 + xyz) + (xz^2 + x^2z + xyz) + (y^2z + yz^2 + xyz)$$
$$- xyz$$
$$= xy(x+y+z) + xz(x+y+z) + yz(x+y+z) - xyz$$
$$= (x+y+z)(xy+yz+zx) - xyz.$$

2. $(x+y+z)^2 = x^2 + y^2 + z^2 + 2(xy+yz+zx)$
   We have

$$(x+y+z)^2 = [(x+y)+z]^2$$
$$= (x+y)^2 + 2z(x+y) + z^2$$
$$= x^2 + 2xy + y^2 + 2zx + 2yz + z^2$$
$$= x^2 + y^2 + z^2 + 2(xy+yz+zx).$$

3. $x^3+y^3+z^3-3xyz = (x+y+z)(x^2+y^2+z^2-xy-yz-zx)$
   We have

$$x^3 + y^3 + z^3 - 3xyz$$
$$= x^3 + 3x^2y + 3xy^2 + y^3 + z^3 - 3x^2y - 3xy^2 - 3xyz$$
$$= (x+y)^3 + z^3 - 3xy(x+y+z)$$
$$= (x+y+z)\left[(x+y)^2 - z(x+y) + z^2\right] - 3xy(x+y+z)$$
$$= (x+y+z)(x^2 + 2xy + y^2 - xz - yz + z^2 - 3xy)$$
$$= (x+y+z)(x^2 + y^2 + z^2 - xy - yz - zx).$$

Problem 5. Given three real numbers a, b and c such that $\frac{1}{a} + \frac{1}{b} + \frac{1}{c} = 0$ and $a+b+c = 1$. Compute $a^2 + b^2 + c^2$.

Solution. Compute $a^2 + b^2 + c^2$.
We have $\dfrac{1}{a} + \dfrac{1}{b} + \dfrac{1}{c} = 0$.
It follows that
$$\frac{ab + bc + ca}{abc} = 0$$
or
$$ab + bc + ca = 0.$$
From Problem 4(2), we obtain
$$(a + b + c)^2 = a^2 + b^2 + c^2 + 2(ab + bc + ca).$$
Since $a + b + c = 1$, it implies that $1^2 = a^2 + b^2 + c^2 + 2(0)$.
Therefore, $a^2 + b^2 + c^2 = 1$.

Problem 6. Let a, b and c be three distinct rational numbers. Prove that
$$\frac{1}{(a-b)^2} + \frac{1}{(b-c)^2} + \frac{1}{(c-a)^2}$$
is a square of a rational number.

Solution. Prove that $\dfrac{1}{(a-b)^2} + \dfrac{1}{(b-c)^2} + \dfrac{1}{(c-a)^2}$ is a square of a rational number.
Observe that
$$\frac{1}{(a-b)(b-c)} + \frac{1}{(b-c)(c-a)} + \frac{1}{(c-a)(a-b)}$$
$$= \frac{c-a+a-b+b-c}{(a-b)(b-c)(c-a)} = 0.$$

We obtain
$$\frac{1}{(a-b)^2} + \frac{1}{(b-c)^2} + \frac{1}{(c-a)^2}$$
$$= \frac{1}{(a-b)^2} + \frac{1}{(b-c)^2} + \frac{1}{(c-a)^2}$$
$$+ 2\left[\frac{1}{(a-b)(b-c)} + \frac{1}{(b-c)(c-a)} + \frac{1}{(c-a)(a-b)}\right]$$
$$= \left(\frac{1}{a-b} + \frac{1}{b-c} + \frac{1}{c-a}\right)^2.$$

Since a, b and c are rational numbers, $\dfrac{1}{a-b} + \dfrac{1}{b-c} + \dfrac{1}{c-a}$ is also a rational number.

Therefore, $\dfrac{1}{(a-b)^2} + \dfrac{1}{(b-c)^2} + \dfrac{1}{(c-a)^2}$ is a square of a rational number.

Problem 7. Let a, b and c be three real number such that $a+b+c = 0$. Prove that
$$\left(\dfrac{a-b}{c} + \dfrac{b-c}{a} + \dfrac{c-a}{b}\right)\left(\dfrac{c}{a-b} + \dfrac{a}{b-c} + \dfrac{b}{c-a}\right) = 9.$$

Solution. Prove that
$$\left(\dfrac{a-b}{c} + \dfrac{b-c}{a} + \dfrac{c-a}{b}\right)\left(\dfrac{c}{a-b} + \dfrac{a}{b-c} + \dfrac{b}{c-a}\right) = 9.$$

We have
$$\begin{aligned}
P &= \dfrac{a-b}{c} + \dfrac{b-c}{a} + \dfrac{c-a}{b} \\
&= \dfrac{ab(a-b) + bc(b-c) + ca(c-a)}{abc} \\
&= \dfrac{ab(a-b) + b^2c - bc^2 + ac^2 - a^2c}{abc} \\
&= \dfrac{ab(a-b) + (b^2c - a^2c) + (-bc^2 + ac^2)}{abc} \\
&= \dfrac{ab(a-b) - c(a^2 - b^2) + c^2(a-b)}{abc} \\
&= \dfrac{ab(a-b) - c(a-b)(a+b) + c^2(a-b)}{abc} \\
&= \dfrac{(a-b)(ab - ac - bc + c^2)}{abc} \\
&= \dfrac{(a-b)[a(b-c) - c(b-c)]}{abc} \\
&= \dfrac{(a-b)(b-c)(a-c)}{abc}.
\end{aligned}$$

Moreover, $Q = \dfrac{c}{a-b} + \dfrac{a}{b-c} + \dfrac{b}{c-a}$.

Let $x = a - b, y = b - c$ and $z = c - a$. Since $a + b + c = 0$, it implies that
$$x - y = a - b - (b - c)$$

$$= a - b - b + c$$
$$= a + c - 2b = a + b + c - 3b = -3b.$$

Then $b = -\dfrac{x-y}{3}$.

Similarly, $a = -\dfrac{z-x}{3}$ and $c = -\dfrac{y-z}{3}$.

It follows that

$$\begin{aligned}
Q &= -\frac{y-z}{3x} - \frac{z-x}{3y} - \frac{x-y}{3z} \\
&= -\frac{1}{3}\left(\frac{x-y}{z} + \frac{y-z}{x} + \frac{z-x}{y}\right) \\
&= -\frac{(x-y)(y-z)(x-z)}{3xyz} \\
&= \frac{(x-y)(y-z)(z-x)}{3xyz} \\
&= \frac{(-3a)(-3b)(-3c)}{3(a-b)(b-c)(c-a)} \\
&= \frac{9abc}{(a-b)(b-c)(a-c)}.
\end{aligned}$$

We obtain

$$PQ = \frac{(a-b)(b-c)(a-c)}{abc} \times \frac{9abc}{(a-b)(b-c)(a-c)} = 9.$$

Chapter 6

Square Roots

The square root of a number is a value that, when multiplied by itself, gives the original number. It is one of the fundamental operations in mathematics that maths learners must know. We will discuss its definition and properties in detail in the next section. Let us first begin this chapter with a very useful symbol. It is called absolute value.

6.1 Absolute Value

6.1.1 Definition

The absolute value of a real number x is the non-negative value of x regardless to its sign. That is, $|x| = \begin{cases} x, & \text{if } x \geq 0 \\ -x, & \text{if } x < 0 \end{cases}$.

> **Example 32**
>
> Compute the following expressions:
>
> 1. $|2|$;
>
> 2. $|-3|$;
>
> 3. $|3 - 2 - 4|$;
>
> 4. $|-7 + 10 - 3|$.

Solution. Compute the following expressions:

1. $|2| = 2$.

2. $|-3| = 3$.

3. $|3 - 2 - 4| = |-3| = 3$.

4. $|-7 + 10 - 3| = |0| = 0$.

6.1.2 Properties of Absolute Values

For all real numbers a and b, we obtain the following properties:

1. $|a| = |-a|$;

2. $-|a| \leq a \leq |a|$;

3. $|a| = |b|$ if and only if $a = b$ or $a = -b$;

4. $|ab| = |a| \times |b|$;

5. $|a^n| = |a|^n$;

6. $\left|\dfrac{a}{b}\right| = \dfrac{|a|}{|b|}, b \neq 0$;

7. $|a \pm b| \leq |a| + |b|$.

6.2 Square Roots

A square root of a is the number which its square equals a. That is, if x is a square root of a, it follows that $x^2 = a$.

Example 33

- 2 is a square root of 4 because $2^2 = 4$.
- -2 is also a square root of 4 because $(-2)^2 = 4$.

We see that there exists too square roots of 4. They are -2 and 2. 2 is called the positive square root of 4 and represented by $\sqrt{4} = 2$. Additionally, -2 is called the negative square root of 4 and represented by $-\sqrt{4} = -2$.

6.2. Square Roots

Remark 6. Since square of all numbers are always positive, from the definition of square roots, all negative numbers has no square roots.

$\sqrt{...}$ is called radical and the number in the radical is called radicant.

Example 34

What are the square roots of 9?

Solution. Observe that $3^2 = 9$ and $(-3)^2 = 9$.
Consequently, 3 and -3 are the square roots of 9.

Example 35

Compute the following square roots:

1. $\sqrt{16}$;
2. $\sqrt{36}$;
3. $\sqrt{49}$;
4. $\sqrt{64}$;
5. $\sqrt{81}$;
6. $\sqrt{121}$;
7. $\sqrt{100}$;
8. $\sqrt{144}$;
9. $\sqrt{169}$;
10. $\sqrt{196}$;
11. $\sqrt{225}$;
12. $\sqrt{256}$;
13. $\sqrt{625}$.

Solution.
1. $\sqrt{16} = \sqrt{4^2} = 4$.
2. $\sqrt{36} = \sqrt{6^2} = 6$.
3. $\sqrt{49} = \sqrt{7^2} = 7$.
4. $\sqrt{64} = \sqrt{8^2} = 8$.
5. $\sqrt{81} = \sqrt{9^2} = 9$.
6. $\sqrt{121} = \sqrt{11^2} = 11$.
7. $\sqrt{100} = \sqrt{10^2} = 10$.
8. $\sqrt{144} = \sqrt{12^2} = 12$.
9. $\sqrt{169} = \sqrt{13^2} = 13$.
10. $\sqrt{196} = \sqrt{14^2} = 14$.

11. $\sqrt{225} = \sqrt{15^2} = 15$.
12. $\sqrt{256} = \sqrt{16^2} = 16$.
13. $\sqrt{625} = \sqrt{25^2} = 25$.

6.3 Properties of Square Roots

For all positive real numbers a and b, we obtain the following properties:

1. $\sqrt{ab} = \sqrt{a} \times \sqrt{b}$;
2. $\sqrt{\dfrac{a}{b}} = \dfrac{\sqrt{a}}{\sqrt{b}}$, where $b \neq 0$;
3. $\sqrt{a^2 b} = \sqrt{a^2} \times \sqrt{b} = a\sqrt{b}$.

Example 36

Compute the following expressions:

1. $\sqrt{12} - \sqrt{27} + \sqrt{\dfrac{3}{4}}$;
2. $\sqrt{8} - \sqrt{18} + \sqrt{72}$;
3. $\sqrt{20} - 3\sqrt{2} - \sqrt{\dfrac{5}{9}} + \sqrt{50}$;
4. $\left(2\sqrt{2} + 5\sqrt{3}\right)\left(3\sqrt{2} - \sqrt{3}\right)$;
5. $\left(\sqrt{3} - \sqrt{2}\right)^2$;
6. $\left(3\sqrt{2} + \sqrt{3}\right)\left(2\sqrt{3} - \sqrt{2}\right)^2$.

Solution. Compute the following expressions:

1. $\sqrt{12} - \sqrt{27} + \sqrt{\dfrac{3}{4}}$

We have

$$\sqrt{12} - \sqrt{27} + \sqrt{\dfrac{3}{4}} = \sqrt{2^2 \times 3} - \sqrt{3^2 \times 3} + \sqrt{\dfrac{3}{2^2}}$$

6.3. Properties of Square Roots

$$= 2\sqrt{3} - 3\sqrt{3} + \frac{\sqrt{3}}{2}$$
$$= -\sqrt{3} + \frac{\sqrt{3}}{2} = \frac{-2\sqrt{3} + \sqrt{3}}{2}$$
$$= -\frac{\sqrt{3}}{2}.$$

Therefore, $\sqrt{12} - \sqrt{27} + \sqrt{\frac{3}{4}} = -\frac{\sqrt{3}}{2}$.

2. $\sqrt{8} - \sqrt{18} + \sqrt{72}$
   We have

$$\sqrt{8} - \sqrt{18} + \sqrt{72} = \sqrt{2^2 \times 2} - \sqrt{3^2 \times 2} + \sqrt{6^2 \times 2}$$
$$= 2\sqrt{2} - 3\sqrt{2} + 6\sqrt{2} = 5\sqrt{2}.$$

Therefore, $\sqrt{8} - \sqrt{18} + \sqrt{72} = 2\sqrt{2} - 3\sqrt{2} + 6\sqrt{2} = 5\sqrt{2}$.

3. $\sqrt{20} - 3\sqrt{2} - \sqrt{\frac{5}{9}} + \sqrt{50}$
   We have

$$\sqrt{20} - 3\sqrt{2} - \sqrt{\frac{5}{9}} + \sqrt{50} = \sqrt{2^2 \times 5} - 3\sqrt{2} - \sqrt{\frac{5}{3^2}} + \sqrt{5^2 \times 2}$$
$$= 2\sqrt{5} - 3\sqrt{2} - \frac{\sqrt{5}}{3} + 5\sqrt{2}$$
$$= 2\sqrt{5} - \frac{\sqrt{5}}{3} + 5\sqrt{2} - 3\sqrt{2}$$
$$= \frac{6\sqrt{5} - \sqrt{5}}{3} + 2\sqrt{2}$$
$$= \frac{5\sqrt{5}}{3} + 2\sqrt{2}.$$

Therefore, $\sqrt{20} - 3\sqrt{2} - \sqrt{\frac{5}{9}} + \sqrt{50}$.

4. $\left(2\sqrt{2} + 5\sqrt{3}\right)\left(3\sqrt{2} - \sqrt{3}\right)$
   We have

$$\left(2\sqrt{2} + 5\sqrt{3}\right)\left(3\sqrt{2} - \sqrt{3}\right)$$

$$= 6\sqrt{2^2} - 2\sqrt{6} + 15\sqrt{6} - 5\sqrt{3^2}$$
$$= 6(2) + 13\sqrt{6} - 5(3)$$
$$= 12 + 13\sqrt{6} - 15$$
$$= -3 + 13\sqrt{6}.$$

Therefore, $\left(2\sqrt{2} + 5\sqrt{3}\right)\left(3\sqrt{2} - \sqrt{3}\right) = -3 + 13\sqrt{6}.$

5. $\left(\sqrt{3} - \sqrt{2}\right)^2$
We have
$$\left(\sqrt{3} - \sqrt{2}\right)^2 = \sqrt{3^2} - 2\sqrt{3}\sqrt{2} + \sqrt{2^2}$$
$$= 3 - 2\sqrt{6} + 2$$
$$= 5 - 2\sqrt{6}.$$

Therefore, $\left(\sqrt{3} - \sqrt{2}\right)^2 = 5 - 2\sqrt{6}.$

6. $\left(3\sqrt{2} + \sqrt{3}\right)\left(2\sqrt{3} - \sqrt{2}\right)^2$
We have
$$\left(3\sqrt{2} + \sqrt{3}\right)\left(2\sqrt{3} - \sqrt{2}\right)^2$$
$$= \left(3\sqrt{2} + \sqrt{3}\right)\left[\left(2\sqrt{3}\right)^2 - 2\left(2\sqrt{3}\right)\left(\sqrt{2}\right) + \sqrt{2^2}\right]$$
$$= \left(3\sqrt{2} + \sqrt{3}\right)\left(12 - 4\sqrt{6} + 2\right)$$
$$= \left(3\sqrt{2} + \sqrt{3}\right)\left(14 - 4\sqrt{6}\right)$$
$$= 42\sqrt{2} - 12\sqrt{12} + 14\sqrt{3} - 4\sqrt{18}$$
$$= 42\sqrt{2} - 24\sqrt{3} + 14\sqrt{3} - 12\sqrt{2}$$
$$= 30\sqrt{2} - 10\sqrt{3}.$$

Therefore, $\left(3\sqrt{2} + \sqrt{3}\right)\left(2\sqrt{3} - \sqrt{2}\right)^2 = 30\sqrt{2} - 10\sqrt{3}..$

6.4 Comparing Radical

For all positive real numbers a and b, $a > b$ if and only if $a^2 > b^2$. Hence, to prove that $a > b$, it is sufficient to prove that $a^2 > b^2$ or

$a^2 - b^2 > 0$.

> **Example 37**
>
> Compare $2 + \sqrt{6}$ and $2\sqrt{5}$.

Solution. Observe that

$$\left(2 + \sqrt{6}\right)^2 - \left(2\sqrt{5}\right)^2 = 2^2 + 2(2)\left(\sqrt{6}\right) + \sqrt{6}^2 - 2^2\sqrt{5}^2$$
$$= 4 + 4\sqrt{6} + 6 - 20$$
$$= 4\sqrt{6} - 10$$
$$= \sqrt{4^2 \times 6} - \sqrt{10^2}$$
$$= \sqrt{96} - \sqrt{100} < 0.$$

Then $(2 + \sqrt{6})^2 < (2\sqrt{5})^2$.
Therefore, $2 + \sqrt{6} < 2\sqrt{5}$.

6.5 Rationalize the Denominator

Generally, in Mathematics, we never keep fractions with irrational denominators. Namely, we have to rationalize the denominator of such fractions. In this section, we will show the readers about how to rationalize the denominator of a fraction.

6.6 Fraction in Form $\dfrac{?}{\sqrt[n]{a}}$

To rationalize the denominator of fraction in the form of $\dfrac{?}{\sqrt[n]{a}}$, we have to multiply $\sqrt[n]{a^{n-1}}$ to both parts of the fraction, denominator and numerator. See the following examples.

> **Example 38**
>
> Rationalize the following fractions:

Chapter 6. Square Roots

1. $\dfrac{2}{\sqrt{2}}$;

2. $\dfrac{\sqrt{3}-1}{\sqrt{3}}$;

3. $\dfrac{2}{\sqrt[3]{3}}$;

4. $\dfrac{7\sqrt{2}}{\sqrt{3}}$;

5. $\dfrac{4\sqrt[3]{2}}{\sqrt[3]{5}}$.

Solution. Rationalize the denominators of following fractions:

1. $\dfrac{2}{\sqrt{2}}$

   We have $\dfrac{2}{\sqrt{2}} = \dfrac{2 \times \sqrt{2}}{\sqrt{2} \times \sqrt{2}} = \dfrac{2\sqrt{2}}{\sqrt{2^2}} = \dfrac{2\sqrt{2}}{2} = \sqrt{2}.$

   Therefore, $\dfrac{2}{\sqrt{2}} = \sqrt{2}.$

2. $\dfrac{\sqrt{3}-1}{\sqrt{3}}$

   We have $\dfrac{\sqrt{3}-1}{\sqrt{3}} = \dfrac{\left(\sqrt{3}-1\right)\sqrt{3}}{\sqrt{3} \times \sqrt{3}} = \dfrac{\sqrt{3^2}-\sqrt{3}}{\sqrt{3^2}} = \dfrac{3-\sqrt{3}}{3}.$

   Therefore, $\dfrac{\sqrt{3}-1}{\sqrt{3}} = \dfrac{3-\sqrt{3}}{3}.$

3. $\dfrac{2}{\sqrt[3]{3}}.$

   We have $\dfrac{2}{\sqrt[3]{3}} = \dfrac{2 \times \sqrt[3]{3^2}}{\sqrt[3]{3} \times \sqrt[3]{3^2}} = \dfrac{2\sqrt[3]{9}}{\sqrt[3]{3^3}} = \dfrac{2\sqrt[3]{9}}{3}.$

   Therefore, $\dfrac{2}{\sqrt[3]{3}} = \dfrac{2\sqrt[3]{9}}{3}.$

4. $\dfrac{7\sqrt{2}}{\sqrt{3}}.$

   We have $\dfrac{7\sqrt{2}}{\sqrt{3}} = \dfrac{7\sqrt{2} \times \sqrt{3}}{\sqrt{3} \times \sqrt{3}} = \dfrac{7\sqrt{6}}{\sqrt{3^2}} = \dfrac{7\sqrt{6}}{3}.$

   Therefore, $\dfrac{7\sqrt{2}}{\sqrt{3}} = \dfrac{7\sqrt{6}}{3}.$

5. $\dfrac{4\sqrt[3]{2}}{\sqrt[3]{5}}.$

6.6. Fraction in Form $\dfrac{?}{\sqrt[n]{a}}$

We have $\dfrac{4\sqrt[3]{2}}{\sqrt[3]{5}} = \dfrac{4\sqrt[3]{2} \times \sqrt[3]{5^2}}{\sqrt[3]{5} \times \sqrt[3]{5^2}} = \dfrac{4\sqrt[3]{2 \times 25}}{\sqrt[3]{5^3}} = \dfrac{4\sqrt[3]{50}}{5}$.

Therefore, $\dfrac{4\sqrt[3]{2}}{\sqrt[3]{5}} = \dfrac{4\sqrt[3]{50}}{5}$.

6.6.1 Fraction in the Form of $\dfrac{?}{\sqrt{a}+\sqrt{b}}$ or $\dfrac{?}{\sqrt{a}-\sqrt{b}}$

To rationalize the denominators of fractions in the forms of $\dfrac{?}{\sqrt{a}+\sqrt{b}}$ or $\dfrac{?}{\sqrt{a}-\sqrt{b}}$, we use the formula $a^2 - b^2 = (a-b)(a+b)$. That is:

- We multiply $\sqrt{a}-\sqrt{b}$ to both parts of fraction in form $\dfrac{?}{\sqrt{a}+\sqrt{b}}$.

- We multiply $\sqrt{a}+\sqrt{b}$ to both parts of fraction in form $\dfrac{?}{\sqrt{a}-\sqrt{b}}$.

Remark 7. Note that $\sqrt{a}-\sqrt{b}$ is called the conjugate of $\sqrt{a}+\sqrt{b}$.

See the following examples:

Example 39

Rationalize the denominators of the following fractions:

1. $\dfrac{1}{\sqrt{3}+\sqrt{2}}$;

2. $\dfrac{\sqrt{2}}{1-\sqrt{2}}$;

3. $\dfrac{1}{\sqrt{5}-\sqrt{3}}$;

4. $\dfrac{\sqrt{3}}{\sqrt{7}+\sqrt{5}}$;

5. $\dfrac{1}{1+\sqrt{2}-\sqrt{3}}$.

Solution. Rationalize the denominators of the following fractions:

1. $\dfrac{1}{\sqrt{3}+\sqrt{2}}$

We have

$$\dfrac{1}{\sqrt{3}+\sqrt{2}} = \dfrac{\sqrt{3}-\sqrt{2}}{\left(\sqrt{3}+\sqrt{2}\right)\left(\sqrt{3}-\sqrt{2}\right)}$$

$$= \frac{\sqrt{3}-\sqrt{2}}{\sqrt{3^2}-\sqrt{2^2}}$$
$$= \frac{\sqrt{3}-\sqrt{2}}{3-2}$$
$$= \frac{\sqrt{3}-\sqrt{2}}{1}$$
$$= \sqrt{3}-\sqrt{2}.$$

Therefore, $\dfrac{1}{\sqrt{3}+\sqrt{2}} = \sqrt{3}-\sqrt{2}.$

2. $\dfrac{\sqrt{2}}{1-\sqrt{2}}$
We have

$$\frac{\sqrt{2}}{1-\sqrt{2}} = \frac{\sqrt{2}\left(1+\sqrt{2}\right)}{\left(1-\sqrt{2}\right)\left(1+\sqrt{2}\right)}$$
$$= \frac{\sqrt{2}+\sqrt{2^2}}{1^2-\sqrt{2^2}}$$
$$= \frac{\sqrt{2}+2}{1-2}$$
$$= \frac{2+\sqrt{2}}{-1}$$
$$= -2-\sqrt{2}.$$

Therefore, $\dfrac{\sqrt{2}}{1-\sqrt{2}} = -2-\sqrt{2}.$

3. $\dfrac{1}{\sqrt{5}-\sqrt{3}}$
We have

$$\frac{1}{\sqrt{5}-\sqrt{3}} = \frac{\sqrt{5}+\sqrt{3}}{\left(\sqrt{5}-\sqrt{3}\right)\left(\sqrt{5}+\sqrt{3}\right)}$$
$$= \frac{\sqrt{5}+\sqrt{3}}{\sqrt{5^2}-\sqrt{3^2}}$$
$$= \frac{\sqrt{5}+\sqrt{3}}{5-3}$$

6.6. Fraction in Form $\dfrac{?}{\sqrt[n]{a}}$

$$= \frac{\sqrt{5}+\sqrt{3}}{2}.$$

Therefore, $\dfrac{1}{\sqrt{5}-\sqrt{3}} = \dfrac{\sqrt{5}+\sqrt{3}}{2}.$

4. $\dfrac{\sqrt{3}}{\sqrt{7}+\sqrt{5}}$
We have

$$\frac{\sqrt{3}}{\sqrt{7}+\sqrt{5}} = \frac{\sqrt{3}\left(\sqrt{7}-\sqrt{5}\right)}{\left(\sqrt{7}+\sqrt{5}\right)\left(\sqrt{7}-\sqrt{5}\right)}$$
$$= \frac{\sqrt{21}-\sqrt{15}}{\sqrt{7^2}-\sqrt{5^2}}$$
$$= \frac{\sqrt{21}-\sqrt{15}}{7-5}$$
$$= \frac{\sqrt{21}-\sqrt{15}}{2}.$$

Therefore, $\dfrac{\sqrt{3}}{\sqrt{7}+\sqrt{5}} = \dfrac{\sqrt{21}-\sqrt{15}}{2}.$

5. $\dfrac{1}{1+\sqrt{2}-\sqrt{3}}$
We have

$$\frac{1}{1+\sqrt{2}-\sqrt{3}} = \frac{1+\sqrt{2}+\sqrt{3}}{\left(1+\sqrt{2}-\sqrt{3}\right)\left(1+\sqrt{2}+\sqrt{3}\right)}$$
$$= \frac{1+\sqrt{2}+\sqrt{3}}{\left(1+\sqrt{2}\right)^2 - \sqrt{3}^2}$$
$$= \frac{1+\sqrt{2}+\sqrt{3}}{1+2\sqrt{2}+\sqrt{2^2}-3}$$
$$= \frac{1+\sqrt{2}+\sqrt{3}}{1+2\sqrt{2}+2-3}$$
$$= \frac{1+\sqrt{2}+\sqrt{3}}{2\sqrt{2}}$$
$$= \frac{\left(1+\sqrt{2}+\sqrt{3}\right)\sqrt{2}}{2\sqrt{2^2}}$$

$$= \frac{\sqrt{2} + \sqrt{2^2} + \sqrt{6}}{4}$$

$$= \frac{2 + \sqrt{2} + \sqrt{6}}{4}.$$

Therefore, $\dfrac{1}{1 + \sqrt{2} - \sqrt{3}} = \dfrac{2 + \sqrt{2} + \sqrt{6}}{4}.$

6.7 Simplify Expressions in the Form of $\sqrt{a + 2\sqrt{b}}$

To simplify this kind of expressions, we have to write $a + 2\sqrt{b}$ in form $c + d + 2\sqrt{cd} = \sqrt{c^2} + 2\sqrt{cd} + \sqrt{d^2} = \left(\sqrt{c} + \sqrt{d}\right)^2$. That is, we have to find two numbers c and d such that $c + d = a$ and $cd = b$. See the following examples.

Example 40

Simplify the following expressions:

1. $\sqrt{7 + 2\sqrt{10}}$;

2. $\sqrt{5 + 2\sqrt{6}}$;

3. $\sqrt{4 + 2\sqrt{3}}$;

4. $\sqrt{12 - 2\sqrt{20}}$;

5. $\sqrt{8 - 2\sqrt{12}}$.

Solution. Simplify the following expressions:

1. $\sqrt{7 + 2\sqrt{10}}$
We have

$$\sqrt{7 + 2\sqrt{10}} = \sqrt{5 + 2\sqrt{10} + 2}$$
$$= \sqrt{\sqrt{5^2} + 2\sqrt{10} + \sqrt{2^2}}$$
$$= \sqrt{\left(\sqrt{5} + \sqrt{2}\right)^2}$$
$$= \sqrt{5} + \sqrt{2}.$$

6.7. Simplify Expressions in the Form of $\sqrt{a+2\sqrt{b}}$

Therefore, $\sqrt{7+2\sqrt{10}} = \sqrt{5} + \sqrt{2}$.

2. $\sqrt{5+2\sqrt{6}}$
We have
$$\sqrt{5+2\sqrt{6}} = \sqrt{3+2\sqrt{6}+2}$$
$$= \sqrt{\sqrt{3^2}+2\sqrt{6}+\sqrt{2^2}}$$
$$= \sqrt{\left(\sqrt{3}+\sqrt{2}\right)^2}$$
$$= \sqrt{3}+\sqrt{2}.$$

Therefore, $\sqrt{5+2\sqrt{6}} = \sqrt{3}+\sqrt{2}$.

3. $\sqrt{4+2\sqrt{3}}$
We have
$$\sqrt{4+2\sqrt{3}} = \sqrt{3+2\sqrt{3}+1}$$
$$= \sqrt{\sqrt{3^2}+2\sqrt{3}+1^2}$$
$$= \sqrt{\left(\sqrt{3}+1\right)^2}$$
$$= \sqrt{3}+1.$$

Therefore, $\sqrt{4+2\sqrt{3}} = \sqrt{3}+1$.

4. $\sqrt{12-2\sqrt{20}}$
We have
$$\sqrt{12-2\sqrt{20}} = \sqrt{10-2\sqrt{20}+2}$$
$$= \sqrt{\sqrt{10^2}-2\sqrt{20}+\sqrt{2^2}}$$
$$= \sqrt{\left(\sqrt{10}-\sqrt{2}\right)^2}$$
$$= \sqrt{10}-\sqrt{2}.$$

Therefore, $\sqrt{12-2\sqrt{20}} = \sqrt{10}-\sqrt{2}$.

5. $\sqrt{8 - 2\sqrt{12}}$
   We have
   $$\begin{aligned} \sqrt{8 - 2\sqrt{12}} &= \sqrt{6 - 2\sqrt{12} + 2} \\ &= \sqrt{\sqrt{6^2} - 2\sqrt{12} + \sqrt{2^2}} \\ &= \sqrt{\left(\sqrt{6} - \sqrt{2}\right)^2} \\ &= \sqrt{6} - \sqrt{2}. \end{aligned}$$
   Therefore, $\sqrt{8 - 2\sqrt{12}} = \sqrt{6} - \sqrt{2}$.

Exercises

Problem 1. Evaluate the following expressions:

1. $|-2|$;
2. $|5|$;
3. $|-7|$;
4. $|-1|$;
5. $|-3|$;
6. $|-4|$;
7. $|-2+3-1|$;
8. $|-7-8-9|$;
9. $|-2+3-4|$;
10. $|-2^2|$.

Problem 2. Simplify the following expressions:

1. $|3-|-2||$;
2. $|-5-|-1||$;
3. $|4-|-3||$;
4. $|-4-|-2+|-1|||$;
5. $|-3-2|1-2||$.

Problem 3. Simplify the following expressions:

1. $-\sqrt{25}$;
2. $\sqrt{(-4)^2}$;
3. $\sqrt{(-8)^2}$;
4. $\sqrt{64}$;
5. $\sqrt{49}$;
6. $\sqrt[3]{(-6)^3}$;
7. $\sqrt[3]{(-3)^3}$;
8. $\sqrt[3]{(-125)^5}$;

9. $-\sqrt[3]{64}$;

10. $\sqrt[4]{(-4)^2}$.

Problem 4. Simplify the following expressions:

1. $\sqrt{x^2}$;
2. $\sqrt{4x^2}$;
3. $\sqrt{9x^2y^2}$;
4. $\sqrt{(x-3)^2}$;
5. $\sqrt{x^2-2xy+y^2}$;
6. $\sqrt{x^2-6x+9}$;
7. $\sqrt{x^2-4xy+4y^2}$;
8. $\sqrt{\dfrac{9}{x^2}}$;
9. $\sqrt[3]{27x^3y^3}$;
10. $\sqrt[3]{(x-3)^3}$.

Problem 5. Simplify the following expressions:

1. $\sqrt{32x^2y^4}$;
2. $\sqrt{16x^4y^6}$;
3. $\sqrt{24x^3y^3}$;
4. $\sqrt{64x^3y^4}$;
5. $\sqrt{112x^6y^9}$;
6. $\sqrt[3]{8x^3y^6}$;
7. $\sqrt[3]{-27x^6y^7}$;
8. $\sqrt[3]{125x^3y^6}$;
9. $\sqrt[3]{-81x^4y^5}$;
10. $\sqrt[3]{(-3)^3 x^7 y^8}$.

Problem 6. Evaluate the following expressions:

1. $\sqrt{2}+2\sqrt{2}$;
2. $6\sqrt{2}-7\sqrt{2}$;
3. $3\sqrt{2}-4\sqrt{2}$;
4. $3\sqrt{3}-2\sqrt{3}$;
5. $5\sqrt{3}+2\sqrt{3}+4\sqrt{3}$;
6. $7\sqrt{7}-3\sqrt{7}+2\sqrt{7}$;
7. $2\sqrt{6}-3\sqrt{6}-7\sqrt{6}$;
8. $3\left(2\sqrt{2}-3\sqrt{2}\right)-4\left(3\sqrt{2}-4\sqrt{2}\right)$;

6.7. Simplify Expressions in the Form of $\sqrt{a+2\sqrt{b}}$

9. $2\sqrt{8} - 3\sqrt{32}$;
10. $4\sqrt{18} - 3\sqrt{8} + 7\sqrt{50}$.

Problem 7. Evaluate the following expressions:

1. $\sqrt{2} \times \sqrt{3}$;
2. $\sqrt{4} \times \sqrt{8}$;
3. $\sqrt{3} \times \sqrt{5} \times \sqrt{15}$;
4. $\sqrt{6} \times \sqrt{2} \times \sqrt{3}$;
5. $\sqrt{3}\left(-\sqrt{2}+\sqrt{3}\right)$;
6. $\sqrt{5}\left(\sqrt{5}+\sqrt{125}\right)$;
7. $\left(\sqrt{2}+\sqrt{3}\right)\left(\sqrt{2}+\sqrt{27}\right)$;
8. $\left(3\sqrt{2}-2\sqrt{3}\right)\left(4\sqrt{2}+3\sqrt{3}\right)$;
9. $\left(\sqrt{3}-1\right)\left(\sqrt{3}-2\right)\left(\sqrt{3}-3\right)$;
10. $\left(1-\sqrt{2}\right)\left(1-\sqrt{3}\right)\left(1+\sqrt{2}\right)\left(1+\sqrt{3}\right)$.

Problem 8. Evaluate the following expressions:

1. $\left(\sqrt{2}+\sqrt{3}\right)\left(\sqrt{3}+\sqrt{4}\right)$;
2. $\left(1-\sqrt{5}\right)\left(1-\sqrt{125}\right)$;
3. $\left(2\sqrt{2}-\sqrt{3}\right)\left(\sqrt{2}+2\sqrt{3}\right)$;
4. $\left(\sqrt{2}+2\sqrt{3}\right)\left(\sqrt{2}-2\sqrt{3}\right)$;
5. $\left(\sqrt{2}+\sqrt{27}\right)\left(3\sqrt{2}-3\sqrt{12}\right)$;
6. $\left(3\sqrt{2}-4\sqrt{5}\right)\left(3\sqrt{5}+4\sqrt{2}\right)$;
7. $\left(5\sqrt{2}-3\sqrt{3}\right)\left(\sqrt{8}+\sqrt{27}\right)$;

8. $\left(3\sqrt{5} - \sqrt{6}\right)\left(3\sqrt{6} + 4\sqrt{5}\right)$.

Problem 9. Evaluate the following expressions:

1. $\left(2\sqrt[3]{8} - 3\sqrt[3]{27}\right)\left(4 + \sqrt[3]{64}\right)$;

2. $\left(\sqrt[3]{3} + 1\right)\left(\sqrt[3]{9} + 1\right)$;

3. $\left(\sqrt[3]{3} - 1\right)\left(\sqrt[3]{9} + \sqrt[3]{3} + 1\right)$;

4. $\left(\sqrt[3]{25} - \sqrt[3]{27}\right)\left(\sqrt[3]{5} + \sqrt[3]{3}\right)$;

5. $\left(\sqrt[3]{24} - \sqrt[3]{250}\right)\left(\sqrt[3]{24} + \sqrt[3]{250}\right)$;

6. $\left(\sqrt[3]{2} + \sqrt[3]{3}\right)\left(\sqrt[3]{27} + \sqrt[3]{16}\right)$;

7. $\left(2\sqrt[3]{4} + \sqrt[3]{2}\right)\left(2\sqrt[3]{4} - 3\sqrt[3]{2}\right)$;

8. $\left(\sqrt[3]{3} + 1\right)^3$;

9. $\left(\sqrt[3]{3} - 1\right)^3$;

10. $\left(\sqrt[3]{2} - 1\right)^3 + \left(\sqrt[3]{2} + 1\right)^3$.

Problem 10. Comparing the following numbers:

1. $\sqrt{2}$ and $\sqrt[3]{3}$;
2. $\sqrt[3]{3}$ and $\sqrt[4]{4}$;
3. $\sqrt{3} + 1$ and $2\sqrt{2}$;
4. $\sqrt{6} + \sqrt{8}$ and $2\sqrt{7}$.

Problem 11. Simplify the following expressions:

1. $\sqrt{5 - 2\sqrt{6}}$
2. $\sqrt{5 + 2\sqrt{6}}$
3. $\sqrt{5 - 2\sqrt{4}}$
4. $\sqrt{5 + 2\sqrt{4}}$
5. $\sqrt{7 - 2\sqrt{6}}$
6. $\sqrt{7 + 2\sqrt{6}}$

6.7. Simplify Expressions in the Form of $\sqrt{a + 2\sqrt{b}}$

7. $\sqrt{2 + \sqrt{3}}$

8. $\sqrt{2 - \sqrt{3}}$

9. $\sqrt{4 - \sqrt{15}}$

10. $\sqrt{4 + \sqrt{15}}$

Problem 12. Given that $a = \sqrt{2}+\sqrt{3}$ and $b = \sqrt{2}-2\sqrt{3}$. Evaluate the following expressions:

1. $a^2 + b^2$;
2. $a^2 - b^2$;
3. ab;
4. $a^2 b + ab^2$;
5. $(a + b)^2$;
6. $(a - b)^2$;
7. $(a + b)^3$;
8. $(a - b)^3$;
9. $\dfrac{a^3 + a^2 b}{a + b}$;
10. $\dfrac{a^3 - a^2 b}{a - b}$.

Problem 13. Rationalize the following fractions:

1. $\dfrac{1}{\sqrt{7}}$;
2. $\dfrac{1}{2\sqrt{3}}$;
3. $\dfrac{3}{\sqrt{5}}$;
4. $\dfrac{7}{2\sqrt{3}}$;
5. $\dfrac{8}{\sqrt{2}}$;
6. $\dfrac{3}{\sqrt[3]{3}}$;
7. $\dfrac{2}{\sqrt[3]{2}}$;
8. $\dfrac{5}{\sqrt[3]{4}}$;
9. $\dfrac{6}{\sqrt[3]{9}}$;
10. $\dfrac{5}{\sqrt[3]{25}}$;

Problem 14. Rationalize the following fractions:

1. $\dfrac{2}{\sqrt{3} - \sqrt{2}}$;
2. $\dfrac{4}{\sqrt{4} - \sqrt{3}}$;
3. $\dfrac{7}{2\sqrt{3} + \sqrt{2}}$;
4. $\dfrac{8}{2\sqrt{3} - 3\sqrt{2}}$;

5. $\dfrac{1}{5\sqrt{4}+4\sqrt{5}}$;

6. $\dfrac{2}{\sqrt[3]{3}-\sqrt[3]{2}}$;

7. $\dfrac{4}{\sqrt[3]{4}-\sqrt[3]{3}}$;

8. $\dfrac{7}{2\sqrt[3]{3}+\sqrt[3]{2}}$;

9. $\dfrac{8}{2\sqrt[3]{3}-3\sqrt[3]{2}}$;

10. $\dfrac{1}{5\sqrt[3]{4}+4\sqrt[3]{5}}$.

Solutions

Problem 1. Evaluate the following expressions:

1. $|-2|$;
2. $|5|$;
3. $|-7|$;
4. $|-1|$;
5. $|-3|$;
6. $|-4|$;
7. $|-2+3-1|$;
8. $|-7-8-9|$;
9. $|-2+3-4|$;
10. $\left|-2^2\right|$.

Solution. Evaluate the following expressions:

1. $|-2| = 2$
2. $|5| = 5$
3. $|-7| = 7$
4. $|-1| = 1$
5. $|-3| = 3$
6. $|-4| = 4$
7. $|-2+3-1| = |0| = 0$
8. $|-7-8-9| = |-24| = 24$
9. $|-2+3-4| = |-3| = 3$
10. $\left|-2^2\right| = |-4| = 4$

Problem 2. Simplify the following expressions:

Chapter 6. Square Roots

1. $|3 - |-2|| = |3 - 2| = |1| = 1$
2. $|-5 - |-1|| = |-5 - 1| = |-6| = 6$
3. $|4 - |-3|| = |4 - 3| = |1| = 1$
4.
$$|-4 - |-2 + |-1||| = |-4 - |-2 + 1||$$
$$= |-4 - |-1||$$
$$= |-4 - 1|$$
$$= |-5|$$
$$= 5.$$

5. $|-3 - 2|1 - 2|| = |-3 - 2|-1|| = |-3 - 2| = |-5| = 5$

Problem 3. Simplify the following expressions:

1. $-\sqrt{25}$;
2. $\sqrt{(-4)^2}$;
3. $\sqrt{(-8)^2}$;
4. $\sqrt{64}$;
5. $\sqrt{49}$;
6. $\sqrt[3]{(-6)^3}$;
7. $\sqrt[3]{(-3)^3}$;
8. $\sqrt[3]{(-125)^5}$;
9. $-\sqrt[3]{64}$;
10. $\sqrt[4]{(-4)^2}$.

Solution. Simplify the following expressions:

1. $-\sqrt{25} = -\sqrt{5^2} = -5$
2. $\sqrt{(-4)^2} = |-4| = 4$
3. $\sqrt{(-8)^2} = |-8| = 8$
4. $\sqrt{64} = \sqrt{8^2} = 8$
5. $\sqrt{49} = \sqrt{7^2} = 7$
6. $\sqrt[3]{(-6)^3} = -6$

6.7. Simplify Expressions in the Form of $\sqrt{a+2\sqrt{b}}$

7. $\sqrt[3]{(-3)^3} = -3$

8. $\sqrt[3]{(-125)^5} = \sqrt[3]{(-5^3)^5} = \sqrt[3]{(-5^5)^3} = -5^5$

9. $-\sqrt[3]{64} = -\sqrt[3]{4^3} = -4$

10. $\sqrt[4]{(-4)^2} = \sqrt[4]{16} = \sqrt[4]{2^4} = 2$

Problem 4. Simplify the following expressions:

1. $\sqrt{x^2}$;
2. $\sqrt{4x^2}$;
3. $\sqrt{9x^2y^2}$;
4. $\sqrt{(x-3)^2}$;
5. $\sqrt{x^2 - 2xy + y^2}$;
6. $\sqrt{x^2 - 6x + 9}$;
7. $\sqrt{x^2 - 4xy + y^2}$;
8. $\sqrt{\dfrac{9}{x^2}}$;
9. $\sqrt[3]{27x^3y^3}$;
10. $\sqrt[3]{(x-3)^3}$.

Solution. Simplify the following expressions:

1. $\sqrt{x^2} = |x|$

2. $\sqrt{4x^2} = \sqrt{(2x)^2} = |2x| = 2|x|$

3. $\sqrt{9x^2y^2} = \sqrt{(3xy)^2} = |3xy| = 3|xy|$

4. $\sqrt{(x-3)^2} = |x-3|$

5. $\sqrt{x^2 - 2xy + y^2} = \sqrt{(x-y)^2} = |x-y|$

6. $\sqrt{x^2 - 6x + 9} = \sqrt{(x-3)^2} = |x-3|$

7. $\sqrt{x^2 - 4xy + 4y^2} = \sqrt{(x-2y)^2} = |x-2y|$

8. $\sqrt{\dfrac{9}{x^2}} = \sqrt{\left(\dfrac{3}{x}\right)^2} = \left|\dfrac{3}{x}\right| = \dfrac{3}{|x|}$

9. $\sqrt[3]{27x^3y^3} = \sqrt[3]{(3xy)^3} = 3xy$

10. $\sqrt[3]{(x-3)^3} = x-3$

Problem 5. Simplify the following expressions:

1. $\sqrt{32x^2y^4}$;
2. $\sqrt{16x^4y^6}$;
3. $\sqrt{24x^3y^3}$;
4. $\sqrt{64x^3y^4}$;
5. $\sqrt{112x^6y^9}$;
6. $\sqrt[3]{8x^3y^6}$;
7. $\sqrt[3]{-27x^6y^7}$;
8. $\sqrt[3]{125x^3y^6}$;
9. $\sqrt[3]{-81x^4y^5}$;
10. $\sqrt[3]{(-3)^3 x^7 y^8}$.

Solution. Simplify the following expressions:

1. We have

$$\sqrt{32x^2y^4} = \sqrt{2 \times 4^2 x^2 (y^2)^2}$$
$$= \sqrt{2(4xy^2)^2}$$
$$= |4xy^2|\sqrt{2}$$
$$= 4y^2|x|\sqrt{2}.$$

2. $\sqrt{16x^4y^6}$
We have

$$\sqrt{16x^4y^6} = \sqrt{4^2(x^2)^2(y^3)^2}$$
$$= |4x^2y^3|$$
$$= 4x^2y^2|y|.$$

3. $\sqrt{24x^3y^3}$
We have

$$\sqrt{24x^3y^3} = \sqrt{6xy \times 2^2 x^2 y^2}$$
$$= |2xy|\sqrt{6xy}$$
$$= 2xy\sqrt{6xy}.$$

6.7. Simplify Expressions in the Form of $\sqrt{a+2\sqrt{b}}$

4. $\sqrt{64x^3y^4}$
   We have
   $$\begin{aligned}\sqrt{64x^3y^4} &= \sqrt{8^2x^2(y^2)^2 x}\\ &= |8xy^2|\sqrt{x}\\ &= 8xy^2\sqrt{x}.\end{aligned}$$

5. $\sqrt{112x^6y^9}$
   We have
   $$\begin{aligned}\sqrt{112x^6y^9} &= \sqrt{4^2(x^3)^2(y^4)^2 \times 7y}\\ &= |4x^3y^4|\sqrt{7y}\\ &= 4x^2y^4|x|\sqrt{7y}.\end{aligned}$$

6. $\sqrt[3]{8x^3y^6} = \sqrt[3]{2^3x^3(y^2)^3} = 2xy^2$

7. $\sqrt[3]{-27x^6y^7} = \sqrt[3]{(-3)^3(x^2)^3(y^2)^3 y} = -3x^2y^2\sqrt[3]{y}$

8. $\sqrt[3]{125x^3y^6} = \sqrt[3]{5^3x^3(y^2)^3} = 5xy^2$

9. $\sqrt[3]{-81x^4y^5} = \sqrt[3]{3(-3)^3x^3 \times x \times y^3 \times y^2} = -3xy\sqrt[3]{3xy^2}$

10. $\sqrt[3]{(-3)^3x^7y^8} = \sqrt[3]{(-3)^3(x^2)^3x(y^2)^3y^2} = -3x^2y^2\sqrt[3]{xy^2}$

Problem 6. Evaluate the following expressions:

1. $\sqrt{2} + 2\sqrt{2}$;
2. $6\sqrt{2} - 7\sqrt{2}$;
3. $3\sqrt{2} - 4\sqrt{2}$;
4. $3\sqrt{3} - 2\sqrt{3}$;
5. $5\sqrt{3} + 2\sqrt{3} + 4\sqrt{3}$;
6. $7\sqrt{7} - 3\sqrt{7} + 2\sqrt{7}$;
7. $2\sqrt{6} - 3\sqrt{6} - 7\sqrt{6}$;

8. $3\left(2\sqrt{2}-3\sqrt{2}\right)-4\left(3\sqrt{2}-4\sqrt{2}\right)$;
9. $2\sqrt{8}-3\sqrt{32}$;
10. $4\sqrt{18}-3\sqrt{8}+7\sqrt{50}$.

Solution. Evaluate the following expressions:

1. $\sqrt{2}+2\sqrt{2}=3\sqrt{2}$
2. $6\sqrt{2}-7\sqrt{2}=-\sqrt{2}$
3. $3\sqrt{2}-4\sqrt{2}=-\sqrt{2}$
4. $3\sqrt{3}-2\sqrt{3}=\sqrt{3}$
5. $5\sqrt{3}+2\sqrt{3}+4\sqrt{3}=11\sqrt{3}$
6. $7\sqrt{7}-3\sqrt{7}+2\sqrt{7}=6\sqrt{7}$
7. $2\sqrt{6}-3\sqrt{6}-7\sqrt{6}=-8\sqrt{6}$
8.
$$3\left(2\sqrt{2}-3\sqrt{2}\right)-4\left(3\sqrt{2}-4\sqrt{2}\right)$$
$$=3\left(-\sqrt{2}\right)-4\left(-\sqrt{2}\right)$$
$$=-3\sqrt{2}+4\sqrt{2}$$
$$=\sqrt{2}.$$

9.
$$2\sqrt{8}-3\sqrt{32}=2\sqrt{2^2\times 2}-3\sqrt{4^2\times 2}$$
$$=2\left(2\sqrt{2}\right)-3\left(4\sqrt{2}\right)$$
$$=4\sqrt{2}-12\sqrt{2}$$
$$=-8\sqrt{2}.$$

10.
$$4\sqrt{18}-3\sqrt{8}+7\sqrt{50}=4\sqrt{3^2\times 2}-3\sqrt{2^2\times 2}+7\sqrt{5^2\times 2}$$
$$=12\sqrt{2}-6\sqrt{2}+35\sqrt{2}$$
$$=41\sqrt{2}.$$

6.7. Simplify Expressions in the Form of $\sqrt{a + 2\sqrt{b}}$

Problem 7. Evaluate the following expressions:

1. $\sqrt{2} \times \sqrt{3}$;
2. $\sqrt{4} \times \sqrt{8}$;
3. $\sqrt{3} \times \sqrt{5} \times \sqrt{15}$;
4. $\sqrt{6} \times \sqrt{2} \times \sqrt{3}$;
5. $\sqrt{3}\left(-\sqrt{2} + \sqrt{3}\right)$;
6. $\sqrt{5}\left(\sqrt{5} + \sqrt{125}\right)$;
7. $\left(\sqrt{2} + \sqrt{3}\right)\left(\sqrt{2} + \sqrt{27}\right)$;
8. $\left(3\sqrt{2} - 2\sqrt{3}\right)\left(4\sqrt{2} + 3\sqrt{3}\right)$;
9. $\left(\sqrt{3} - 1\right)\left(\sqrt{3} - 2\right)\left(\sqrt{3} - 3\right)$;
10. $\left(1 - \sqrt{2}\right)\left(1 - \sqrt{3}\right)\left(1 + \sqrt{2}\right)\left(1 + \sqrt{3}\right)$.

Solution. Evaluate the following expressions:

1. $\sqrt{2} \times \sqrt{3} = \sqrt{2 \times 3} = \sqrt{6}$
2. $\sqrt{4} \times \sqrt{8} = \sqrt{4^2 \times 2} = 4\sqrt{2}$
3. $\sqrt{3} \times \sqrt{5} \times \sqrt{15} = \sqrt{3 \times 5 \times 15} = \sqrt{15^2} = 15$
4. $\sqrt{6} \times \sqrt{2} \times \sqrt{3} = \sqrt{6 \times 2 \times 3} = \sqrt{6^2} = 6$
5. $\sqrt{3}\left(-\sqrt{2} + \sqrt{3}\right) = -\sqrt{6} + \sqrt{3^2} = -\sqrt{6} + 3$
6.

$$\sqrt{5}\left(\sqrt{5} + \sqrt{125}\right) = \sqrt{5^2} + \sqrt{625}$$
$$= 5 + \sqrt{25^2}$$
$$= 5 + 25$$
$$= 30.$$

7.
$$\left(\sqrt{2}+\sqrt{3}\right)\left(\sqrt{2}+\sqrt{27}\right)$$
$$=\sqrt{2^2}+\sqrt{54}+\sqrt{6}+\sqrt{81}$$
$$=2+\sqrt{3^2\times 6}+\sqrt{6}+\sqrt{9^2}$$
$$=2+3\sqrt{6}+\sqrt{6}+9$$
$$=11+4\sqrt{6}.$$

8.
$$\left(3\sqrt{2}-2\sqrt{3}\right)\left(4\sqrt{2}+3\sqrt{3}\right)$$
$$=12\sqrt{2^2}+9\sqrt{6}-8\sqrt{6}-6\sqrt{3^2}$$
$$=24+\sqrt{6}-18$$
$$=6+\sqrt{6}.$$

9. $\left(\sqrt{3}-1\right)\left(\sqrt{3}-2\right)\left(\sqrt{3}-3\right)$
We have
$$\left(\sqrt{3}-1\right)\left(\sqrt{3}-2\right)\left(\sqrt{3}-3\right)$$
$$=\left(\sqrt{3^2}-2\sqrt{3}-\sqrt{3}+2\right)\left(\sqrt{3}-3\right)$$
$$=\left(3-3\sqrt{3}+2\right)\left(\sqrt{3}-3\right)$$
$$=\left(5-3\sqrt{3}\right)\left(\sqrt{3}-3\right)$$
$$=5\sqrt{3}-15-3\sqrt{3^2}+9\sqrt{3}$$
$$=-15-9+14\sqrt{3}$$
$$=-24+14\sqrt{3}.$$

10. $\left(1-\sqrt{2}\right)\left(1-\sqrt{3}\right)\left(1+\sqrt{2}\right)\left(1+\sqrt{3}\right)$
We have
$$\left(1-\sqrt{2}\right)\left(1-\sqrt{3}\right)\left(1+\sqrt{2}\right)\left(1+\sqrt{3}\right)$$
$$=\left(1-\sqrt{2}\right)\left(1+\sqrt{2}\right)\left(1-\sqrt{3}\right)\left(1+\sqrt{3}\right)$$

$$= \left(1^2 - \sqrt{2^2}\right)\left(1^2 - \sqrt{3^2}\right)$$
$$= (1-2)(1-3)$$
$$= (-1)(-2)$$
$$= 2.$$

Problem 8. Evaluate the following expressions:

1. $\left(\sqrt{2} + \sqrt{3}\right)\left(\sqrt{3} + \sqrt{4}\right)$;
2. $\left(1 - \sqrt{5}\right)\left(1 - \sqrt{125}\right)$;
3. $\left(2\sqrt{2} - \sqrt{3}\right)\left(\sqrt{2} + 2\sqrt{3}\right)$;
4. $\left(\sqrt{2} + 2\sqrt{3}\right)\left(\sqrt{2} - 2\sqrt{3}\right)$;
5. $\left(\sqrt{2} + \sqrt{27}\right)\left(3\sqrt{2} - 3\sqrt{12}\right)$;
6. $\left(3\sqrt{2} - 4\sqrt{5}\right)\left(3\sqrt{5} + 4\sqrt{2}\right)$;
7. $\left(5\sqrt{2} - 3\sqrt{3}\right)\left(\sqrt{8} + \sqrt{27}\right)$;
8. $\left(3\sqrt{5} - \sqrt{6}\right)\left(3\sqrt{6} + 4\sqrt{5}\right)$.

Solution. Evaluate the following expressions:

1. $\left(\sqrt{2} + \sqrt{3}\right)\left(\sqrt{3} + \sqrt{4}\right)$
   We have
   $$\left(\sqrt{2} + \sqrt{3}\right)\left(\sqrt{3} + \sqrt{4}\right) = \left(\sqrt{2} + \sqrt{3}\right)\left(\sqrt{3} + 2\right)$$
   $$= \sqrt{6} + 2\sqrt{2} + \sqrt{3^2} + 2\sqrt{3}$$
   $$= \sqrt{6} + 2\sqrt{2} + 3 + 2\sqrt{3}$$
   $$= \sqrt{6} + 2\sqrt{3} + 2\sqrt{2} + 3.$$

2. $\left(1 - \sqrt{5}\right)\left(1 - \sqrt{125}\right)$
   We have
   $$\left(1 - \sqrt{5}\right)\left(1 - \sqrt{125}\right) = \left(1 - \sqrt{5}\right)\left(1 - 5\sqrt{5}\right)$$

$$= 1 - 5\sqrt{5} - \sqrt{5} + 5\sqrt{5^2}$$
$$= 1 - 6\sqrt{5} + 25$$
$$= 26 - 6\sqrt{5}.$$

3. $\left(2\sqrt{2} - \sqrt{3}\right)\left(\sqrt{2} + 2\sqrt{3}\right)$
We have
$$\left(2\sqrt{2} - \sqrt{3}\right)\left(\sqrt{2} + 2\sqrt{3}\right) = 2\sqrt{2^2} + 4\sqrt{6} - \sqrt{6} - 2\sqrt{3^2}$$
$$= 4 + 3\sqrt{6} - 6$$
$$= -2 + 3\sqrt{6}.$$

4. $\left(\sqrt{2} + 2\sqrt{3}\right)\left(\sqrt{2} - 2\sqrt{3}\right)$
We have
$$\left(\sqrt{2} + 2\sqrt{3}\right)\left(\sqrt{2} - 2\sqrt{3}\right) = \sqrt{2^2} - 2\sqrt{6} + 2\sqrt{6} - 4\sqrt{3^2}$$
$$= 2 - 12$$
$$= -10.$$

5. $\left(\sqrt{2} + \sqrt{27}\right)\left(3\sqrt{2} - 3\sqrt{12}\right)$
We have
$$\left(\sqrt{2} + \sqrt{27}\right)\left(3\sqrt{2} - 3\sqrt{12}\right) = \left(\sqrt{2} + 3\sqrt{3}\right)\left(3\sqrt{2} - 6\sqrt{3}\right)$$
$$= 3\sqrt{2^2} - 6\sqrt{6} + 9\sqrt{6} - 18\sqrt{3^2}$$
$$= 6 + 3\sqrt{6} - 54$$
$$= -48 + 3\sqrt{6}.$$

6. $\left(3\sqrt{2} - 4\sqrt{5}\right)\left(3\sqrt{5} + 4\sqrt{2}\right)$
We have
$$\left(3\sqrt{2} - 4\sqrt{5}\right)\left(3\sqrt{5} + 4\sqrt{2}\right) = 9\sqrt{10} + 12\sqrt{2^2} - 12\sqrt{5^2} - 16\sqrt{10}$$
$$= 24 - 60 - 7\sqrt{10}$$
$$= -36 - 7\sqrt{10}.$$

6.7. Simplify Expressions in the Form of $\sqrt{a+2\sqrt{b}}$

7. $\left(5\sqrt{2}-3\sqrt{3}\right)\left(\sqrt{8}+\sqrt{27}\right)$

We have
$$\left(5\sqrt{2}-3\sqrt{3}\right)\left(\sqrt{8}+\sqrt{27}\right) = \left(5\sqrt{2}-3\sqrt{3}\right)\left(2\sqrt{2}+3\sqrt{3}\right)$$
$$= 10\sqrt{2^2}+15\sqrt{6}-6\sqrt{6}-3\sqrt{3^2}$$
$$= 20-9+9\sqrt{6}$$
$$= 11+9\sqrt{6}.$$

8. $\left(3\sqrt{5}-\sqrt{6}\right)\left(3\sqrt{6}+4\sqrt{5}\right)$

We have
$$\left(3\sqrt{5}-\sqrt{6}\right)\left(3\sqrt{6}+4\sqrt{5}\right)$$
$$= 9\sqrt{30}+12\sqrt{5^2}-3\sqrt{6^2}-4\sqrt{30}$$
$$= 60-18+5\sqrt{30}$$
$$= 42+5\sqrt{30}.$$

Problem 9. Evaluate the following expressions:

1. $\left(2\sqrt[3]{8}-3\sqrt[3]{27}\right)\left(4+\sqrt[3]{64}\right)$;
2. $\left(\sqrt[3]{3}+1\right)\left(\sqrt[3]{9}+1\right)$;
3. $\left(\sqrt[3]{3}-1\right)\left(\sqrt[3]{9}+\sqrt[3]{3}+1\right)$;
4. $\left(\sqrt[3]{25}-\sqrt[3]{27}\right)\left(\sqrt[3]{5}+\sqrt[3]{3}\right)$;
5. $\left(\sqrt[3]{24}-\sqrt[3]{250}\right)\left(\sqrt[3]{24}+\sqrt[3]{250}\right)$;
6. $\left(\sqrt[3]{2}+\sqrt[3]{3}\right)\left(\sqrt[3]{27}+\sqrt[3]{16}\right)$;
7. $\left(2\sqrt[3]{4}+\sqrt[3]{2}\right)\left(2\sqrt[3]{4}-3\sqrt[3]{2}\right)$;
8. $\left(\sqrt[3]{3}+1\right)^3$;
9. $\left(\sqrt[3]{3}-1\right)^3$;

10. $\left(\sqrt[3]{2}-1\right)^3+\left(\sqrt[3]{2}+1\right)^3$.

Solution. Evaluate the following expressions:

1. $\left(2\sqrt[3]{8}-3\sqrt[3]{27}\right)\left(4+\sqrt[3]{64}\right)$
   We have
   $$\left(2\sqrt[3]{8}-3\sqrt[3]{27}\right)\left(4+\sqrt[3]{64}\right)$$
   $$=\left(2\sqrt[3]{2^3}-3\sqrt[3]{3^3}\right)\left(4+\sqrt[3]{4^3}\right)$$
   $$=(4-9)(4+4)$$
   $$=(-5)(8)$$
   $$=-40.$$

   Therefore, $\left(2\sqrt[3]{8}-3\sqrt[3]{27}\right)\left(4+\sqrt[3]{64}\right)=-40$.

2. $\left(\sqrt[3]{3}+1\right)\left(\sqrt[3]{9}+1\right)$
   We have
   $$\left(\sqrt[3]{3}+1\right)\left(\sqrt[3]{9}+1\right)=\sqrt[3]{27}+\sqrt[3]{3}+\sqrt[3]{9}+1$$
   $$=3+\sqrt[3]{3}+\sqrt[3]{9}+1$$
   $$=\sqrt[3]{9}+\sqrt[3]{3}+4.$$

   Therefore, $\left(\sqrt[3]{3}+1\right)\left(\sqrt[3]{9}+1\right)=\sqrt[3]{9}+\sqrt[3]{3}+4$.

3. $\left(\sqrt[3]{3}-1\right)\left(\sqrt[3]{9}+\sqrt[3]{3}+1\right)$
   We have
   $$\left(\sqrt[3]{3}-1\right)\left(\sqrt[3]{9}+\sqrt[3]{3}+1\right)=\sqrt[3]{3^3}-1^3$$
   $$=3-1$$
   $$=2.$$

   Therefore, $\left(\sqrt[3]{3}-1\right)\left(\sqrt[3]{9}+\sqrt[3]{3}+1\right)=2$.

4. $\left(\sqrt[3]{25}-\sqrt[3]{27}\right)\left(\sqrt[3]{5}+\sqrt[3]{3}\right)$
   We have
   $$\left(\sqrt[3]{25}-\sqrt[3]{27}\right)\left(\sqrt[3]{5}+\sqrt[3]{3}\right)$$

6.7. Simplify Expressions in the Form of $\sqrt{a + 2\sqrt{b}}$

$$\begin{aligned}
&= \left(\sqrt[3]{25} - 3\right)\left(\sqrt[3]{5} + \sqrt[3]{3}\right) \\
&= \sqrt[3]{125} + \sqrt[3]{75} - 3\sqrt[3]{5} - 3\sqrt[3]{3} \\
&= 5 + \sqrt[3]{75} - 3\sqrt[3]{5} - 3\sqrt[3]{3} \\
&= \sqrt[3]{75} - 3\sqrt[3]{5} - 3\sqrt[3]{3} + 5.
\end{aligned}$$

Therefore,

$$\left(\sqrt[3]{25} - \sqrt[3]{27}\right)\left(\sqrt[3]{5} + \sqrt[3]{3}\right) = \sqrt[3]{75} - 3\sqrt[3]{5} - 3\sqrt[3]{3} + 5.$$

5. $\left(\sqrt[3]{24} - \sqrt[3]{250}\right)\left(\sqrt[3]{24} + \sqrt[3]{250}\right)$
We have

$$\begin{aligned}
&\left(\sqrt[3]{24} - \sqrt[3]{250}\right)\left(\sqrt[3]{24} + \sqrt[3]{250}\right) \\
&= \left(2\sqrt[3]{3} - 5\sqrt[3]{2}\right)\left(2\sqrt[3]{3} + 5\sqrt[3]{2}\right) \\
&= \left(2\sqrt[3]{3}\right)^2 - \left(5\sqrt[3]{2}\right)^2 \\
&= 4\sqrt[3]{9} - 25\sqrt[3]{4}.
\end{aligned}$$

Therefore, $\left(\sqrt[3]{24} - \sqrt[3]{250}\right)\left(\sqrt[3]{24} + \sqrt[3]{250}\right) = 4\sqrt[3]{9} - 25\sqrt[3]{4}.$

6. $\left(\sqrt[3]{2} + \sqrt[3]{3}\right)\left(\sqrt[3]{27} + \sqrt[3]{16}\right)$
We have

$$\begin{aligned}
&\left(\sqrt[3]{2} + \sqrt[3]{3}\right)\left(\sqrt[3]{27} + \sqrt[3]{16}\right) \\
&= \left(\sqrt[3]{2} + \sqrt[3]{3}\right)\left(3 + 2\sqrt[3]{2}\right) \\
&= 3\sqrt[3]{2} + 2\sqrt[3]{4} + 3\sqrt[3]{3} + 2\sqrt[3]{6} \\
&= 3\sqrt[3]{2} + 3\sqrt[3]{3} + 2\sqrt[3]{4} + 2\sqrt[3]{6}.
\end{aligned}$$

Therefore,

$$\left(\sqrt[3]{2} + \sqrt[3]{3}\right)\left(\sqrt[3]{27} + \sqrt[3]{16}\right) = 3\sqrt[3]{2} + 3\sqrt[3]{3} + 2\sqrt[3]{4} + 2\sqrt[3]{6}.$$

7. $\left(2\sqrt[3]{4} + \sqrt[3]{2}\right)\left(2\sqrt[3]{4} - 3\sqrt[3]{2}\right)$
We have

$$\left(2\sqrt[3]{4} + \sqrt[3]{2}\right)\left(2\sqrt[3]{4} - 3\sqrt[3]{2}\right)$$

$$= 4\sqrt[3]{16} - 6\sqrt[3]{8} + 2\sqrt[3]{8} - 3\sqrt[3]{4}$$
$$= 8\sqrt[3]{2} - 12 + 4 - 3\sqrt[3]{4}$$
$$= 8\sqrt[3]{2} - 3\sqrt[3]{4} - 8.$$

Therefore,
$$\left(2\sqrt[3]{4} + \sqrt[3]{2}\right)\left(2\sqrt[3]{4} - 3\sqrt[3]{2}\right) = 8\sqrt[3]{2} - 3\sqrt[3]{4} - 8.$$

8. $\left(\sqrt[3]{3} + 1\right)^3$
We have
$$\left(\sqrt[3]{3} + 1\right)^3 = \sqrt[3]{3^3} + 3\sqrt[3]{3^2} + 3\sqrt[3]{3} + 1^3$$
$$= 3 + 3\sqrt[3]{9} + 3\sqrt[3]{3} + 1$$
$$= 3\sqrt[3]{3} + 3\sqrt[3]{9} + 4.$$

Therefore,
$$\left(\sqrt[3]{3} + 1\right)^3 = 3\sqrt[3]{3} + 3\sqrt[3]{9} + 4.$$

9. $\left(\sqrt[3]{3} - 1\right)^3$
We have
$$\left(\sqrt[3]{3} + 1\right)^3 = \sqrt[3]{3^3} - 3\sqrt[3]{3^2} + 3\sqrt[3]{3} - 1^3$$
$$= 3 - 3\sqrt[3]{9} + 3\sqrt[3]{3} - 1$$
$$= 3\sqrt[3]{3} - 3\sqrt[3]{9} + 2.$$

Therefore,
$$\left(\sqrt[3]{3} + 1\right)^3 = 3\sqrt[3]{3} - 3\sqrt[3]{9} + 2.$$

10. $\left(\sqrt[3]{2} - 1\right)^3 + \left(\sqrt[3]{2} + 1\right)^3$
We have
$$\left(\sqrt[3]{2} - 1\right)^3 + \left(\sqrt[3]{2} + 1\right)^3$$
$$= \sqrt[3]{2^3} - 3\sqrt[3]{2^2} + 3\sqrt[3]{2} - 1^3 + \sqrt[3]{2^3} + 3\sqrt[3]{2^2} + 3\sqrt[3]{2} + 1^3$$
$$= 2 + 3\sqrt[3]{2} + 2 + 3\sqrt[3]{2}$$

6.7. Simplify Expressions in the Form of $\sqrt{a+2\sqrt{b}}$

$$= 4 + 6\sqrt[3]{2}.$$

Therefore, $\left(\sqrt[3]{2}-1\right)^3 + \left(\sqrt[3]{2}+1\right)^3 = 4 + 6\sqrt[3]{2}.$

Problem 10. Comparing the following numbers:

1. $\sqrt{2}$ and $\sqrt[3]{3}$;
2. $\sqrt[3]{3}$ and $\sqrt[4]{4}$;
3. $\sqrt{3}+1$ and $2\sqrt{2}$;
4. $\sqrt{6}+\sqrt{8}$ and $2\sqrt{7}$.

Solution. Comparing the following numbers:

1. $\sqrt{2}$ and $\sqrt[3]{3}$
   Observe that
   $$\left(\sqrt{2}\right)^6 - \left(\sqrt[3]{3}\right)^6 = 2^3 - 3^2$$
   $$= 8 - 9$$
   $$= -1 < 0.$$

   It implies that
   $$\left(\sqrt{2}^6\right) < \left(\sqrt[3]{3}\right)^6.$$

   Therefore, $\sqrt{2} < \sqrt[3]{3}.$

2. $\sqrt[3]{3}$ and $\sqrt[4]{4}$
   Observe that
   $$\left(\sqrt[3]{3}\right)^{12} - \left(\sqrt[4]{4}\right)^{12} = 3^4 - 4^3$$
   $$= 81 - 64$$
   $$= 17 > 0.$$

   It follows that
   $$\left(\sqrt[3]{3}\right)^{12} > \left(\sqrt[4]{4}\right)^{12}.$$

   Therefore, $\sqrt[3]{3} > \sqrt[4]{4}.$

3. $\sqrt{3}+1$ and $2\sqrt{2}$
   We have
   $$\left(\sqrt{3}+1\right)^2 - \left(2\sqrt{2}\right)^2$$

$$= \sqrt{3^2} + 2\sqrt{3} + 1^2 - 8$$
$$= 3 + 2\sqrt{3} + 1 - 8$$
$$= 2\sqrt{3} - 4$$
$$= 2\left(\sqrt{3} - 2\right)$$
$$= 2\left(\sqrt{3} - \sqrt{4}\right) < 0.$$

It follows that
$$\left(\sqrt{3} + 1\right)^2 < \left(2\sqrt{2}\right)^2.$$

Therefore,
$$\sqrt{3} + 1 < 2\sqrt{2}.$$

4. $\sqrt{6} + \sqrt{8}$ and $2\sqrt{7}$

We have
$$\left(\sqrt{6} + \sqrt{8}\right)^2 - \left(2\sqrt{7}\right)^2$$
$$= \sqrt{6^2} + 2\sqrt{48} + \sqrt{8^2} - 28$$
$$= 6 + 2\sqrt{48} + 8 - 28$$
$$= 2\sqrt{48} - 14$$
$$= 2\left(\sqrt{48} - 7\right)$$
$$= 2\left(\sqrt{48} - \sqrt{49}\right) < 0.$$

It follows that
$$\left(\sqrt{6} + \sqrt{8}\right)^2 < \left(2\sqrt{7}\right)^2.$$

Therefore, $\sqrt{6} + \sqrt{8} < 2\sqrt{7}$.

Problem 11. Simplify the following expressions:

1. $\sqrt{5 - 2\sqrt{6}}$

2. $\sqrt{5 + 2\sqrt{6}}$

3. $\sqrt{5 - 2\sqrt{4}}$

4. $\sqrt{5 + 2\sqrt{4}}$

5. $\sqrt{7 - 2\sqrt{6}}$

6. $\sqrt{7 + 2\sqrt{6}}$

6.7. Simplify Expressions in the Form of $\sqrt{a + 2\sqrt{b}}$

7. $\sqrt{2 + \sqrt{3}}$

8. $\sqrt{2 - \sqrt{3}}$

9. $\sqrt{4 - \sqrt{15}}$

10. $\sqrt{4 + \sqrt{15}}$

Solution. Simplify the following expressions:

1. $\sqrt{5 - 2\sqrt{6}}$
   We have
   $$\sqrt{5 - 2\sqrt{6}} = \sqrt{3 - 2\sqrt{6} + 2}$$
   $$= \sqrt{\sqrt{3^2} - 2\sqrt{3} \times \sqrt{2} + \sqrt{2^2}}$$
   $$= \sqrt{\left(\sqrt{3} - \sqrt{2}\right)^2}$$
   $$= \sqrt{3} - \sqrt{2}.$$

2. $\sqrt{5 + 2\sqrt{6}}$
   We have
   $$\sqrt{5 + 2\sqrt{6}} = \sqrt{3 + 2\sqrt{6} + 2}$$
   $$= \sqrt{\sqrt{3^2} + 2\sqrt{3} \times \sqrt{2} + \sqrt{2^2}}$$
   $$= \sqrt{\left(\sqrt{3} + \sqrt{2}\right)^2}$$
   $$= \sqrt{3} + \sqrt{2}.$$

3. $\sqrt{5 - 2\sqrt{4}}$
   We have $\sqrt{5 - 2\sqrt{4}} = \sqrt{5 - 4} = \sqrt{1} = 1.$

4. $\sqrt{5 + 2\sqrt{4}}$
   We have $\sqrt{5 + 2\sqrt{4}} = \sqrt{5 + 4} = \sqrt{9} = \sqrt{3^2} = 3.$

5. $\sqrt{7 - 2\sqrt{6}}$
   We have
   $$\sqrt{7 - 2\sqrt{6}} = \sqrt{6 - 2\sqrt{6} + 1}$$

$$= \sqrt{\sqrt{6}^2 - 2\sqrt{6} + 1^2}$$
$$= \sqrt{\left(\sqrt{6} - 1\right)^2}$$
$$= \sqrt{6} - 1.$$

6. $\sqrt{7 + 2\sqrt{6}}$
   We have
$$\sqrt{7 + 2\sqrt{6}} = \sqrt{6 + 2\sqrt{6} + 1}$$
$$= \sqrt{\sqrt{6}^2 - 2\sqrt{6} + 1^2}$$
$$= \sqrt{\left(\sqrt{6} + 1\right)^2}$$
$$= \sqrt{6} + 1.$$

7. $\sqrt{2 + \sqrt{3}}$
   We have
$$\sqrt{2 + \sqrt{3}} = \sqrt{\frac{4 + 2\sqrt{3}}{2}}$$
$$= \frac{\sqrt{4 + 2\sqrt{3}}}{\sqrt{2}}$$
$$= \frac{\sqrt{3 + 2\sqrt{3} + 1}}{\sqrt{2}}$$
$$= \frac{\sqrt{\sqrt{3}^2 + 2\sqrt{3} + 1^2}}{\sqrt{2}}$$
$$= \frac{\sqrt{\left(\sqrt{3} + 1\right)^2}}{\sqrt{2}}$$
$$= \frac{\sqrt{3} + 1}{\sqrt{2}}$$
$$= \frac{\sqrt{6} + \sqrt{2}}{\sqrt{2^2}}$$
$$= \frac{\sqrt{6} + \sqrt{2}}{2}.$$

6.7. Simplify Expressions in the Form of $\sqrt{a + 2\sqrt{b}}$

8. $\sqrt{2 - \sqrt{3}}$
   We have
$$\sqrt{2 - \sqrt{3}} = \sqrt{\frac{4 - 2\sqrt{3}}{2}}$$
$$= \frac{\sqrt{4 - 2\sqrt{3}}}{\sqrt{2}}$$
$$= \frac{\sqrt{3 - 2\sqrt{3} + 1}}{\sqrt{2}}$$
$$= \frac{\sqrt{\sqrt{3}^2 - 2\sqrt{3} + 1^2}}{\sqrt{2}}$$
$$= \frac{\sqrt{\left(\sqrt{3} - 1\right)^2}}{\sqrt{2}}$$
$$= \frac{\sqrt{3} - 1}{\sqrt{2}}$$
$$= \frac{\sqrt{6} - \sqrt{2}}{\sqrt{2^2}}$$
$$= \frac{\sqrt{6} - \sqrt{2}}{2}.$$

9. $\sqrt{4 - \sqrt{15}}$
   We have
$$\sqrt{4 - \sqrt{15}} = \sqrt{\frac{8 - 2\sqrt{15}}{2}}$$
$$= \frac{\sqrt{8 - 2\sqrt{3}}}{\sqrt{2}}$$
$$= \frac{\sqrt{5 - 2\sqrt{3} + 3}}{\sqrt{2}}$$
$$= \frac{\sqrt{\sqrt{5}^2 - 2\sqrt{3} + \sqrt{3}^2}}{\sqrt{2}}$$
$$= \frac{\sqrt{\left(\sqrt{5} - \sqrt{3}\right)^2}}{\sqrt{2}}$$

$$= \frac{\sqrt{5}-\sqrt{3}}{\sqrt{2}}$$
$$= \frac{\sqrt{10}-\sqrt{6}}{\sqrt{2^2}}$$
$$= \frac{\sqrt{10}-\sqrt{6}}{2}.$$

10. $\sqrt{4+\sqrt{15}}$

We have

$$\sqrt{4+\sqrt{15}} = \sqrt{\frac{8+2\sqrt{15}}{2}}$$
$$= \frac{\sqrt{8+2\sqrt{3}}}{\sqrt{2}}$$
$$= \frac{\sqrt{5+2\sqrt{3}+3}}{\sqrt{2}}$$
$$= \frac{\sqrt{\sqrt{5}^2 + 2\sqrt{3} + \sqrt{3}^2}}{\sqrt{2}}$$
$$= \frac{\sqrt{(\sqrt{5}+\sqrt{3})^2}}{\sqrt{2}}$$
$$= \frac{\sqrt{5}+\sqrt{3}}{\sqrt{2}}$$
$$= \frac{\sqrt{10}+\sqrt{6}}{\sqrt{2^2}}$$
$$= \frac{\sqrt{10}+\sqrt{6}}{2}.$$

Problem 12. Given that $a = \sqrt{2}+\sqrt{3}$ and $b = \sqrt{2}-2\sqrt{3}$. Evaluate the following expressions:

1. $a^2 + b^2$;
2. $a^2 - b^2$;
3. ab;
4. $a^2b + ab^2$;
5. $(a+b)^2$;
6. $(a-b)^2$;

6.7. Simplify Expressions in the Form of $\sqrt{a + 2\sqrt{b}}$

7. $(a+b)^3$;

8. $(a-b)^3$;

9. $\dfrac{a^3 + a^2 b}{a + b}$;

10. $\dfrac{a^3 - a^2 b}{a - b}$.

Solution. Evaluate the following expressions:

1. $a^2 + b^2$

Since $a = \sqrt{2} + \sqrt{3}$ and $b = \sqrt{2} - 2\sqrt{3}$, it follows that

$$a^2 + b^2 = \left(\sqrt{2} + \sqrt{3}\right)^2 + \left(\sqrt{2} - 2\sqrt{3}\right)^2$$
$$= \sqrt{2^2} + 2\sqrt{6} + \sqrt{3^2} + \sqrt{2^2} - 4\sqrt{6} + 2^2\sqrt{3^2}$$
$$= 2 + 2\sqrt{6} + 3 + 2 - 4\sqrt{6} + 12$$
$$= 19 - 2\sqrt{6}.$$

2. $a^2 - b^2$

We have

$$a^2 - b^2 = \left(\sqrt{2} + \sqrt{3}\right)^2 - \left(\sqrt{2} - 2\sqrt{3}\right)^2$$
$$= \left(\sqrt{2^2} + 2\sqrt{6} + \sqrt{3^2}\right) - \left(\sqrt{2^2} - 4\sqrt{6} + 2^2\sqrt{3^2}\right)$$
$$= 2 + 2\sqrt{6} + 3 - 2 + 4\sqrt{6} - 12$$
$$= -9 + 6\sqrt{6}.$$

3. ab

We have

$$ab = \left(\sqrt{2} + \sqrt{3}\right)\left(\sqrt{2} - 2\sqrt{3}\right)$$
$$= \sqrt{2^2} - 2\sqrt{6} + \sqrt{6} - 2\sqrt{3^2}$$
$$= 2 - \sqrt{6} - 6$$
$$= -4 - \sqrt{6}.$$

4. $a^2 b + ab^2$

We have

$$a^2 b + ab^2 = ab\,(a + b)$$

$$= \left(-4-\sqrt{6}\right)\left(\sqrt{2}+\sqrt{3}+\sqrt{2}-2\sqrt{3}\right)$$
$$= \left(-4-\sqrt{6}\right)\left(2\sqrt{2}-\sqrt{3}\right)$$
$$= -8\sqrt{2}+4\sqrt{3}-2\sqrt{12}+\sqrt{18}$$
$$= -8\sqrt{2}+4\sqrt{3}-4\sqrt{3}+3\sqrt{2}$$
$$= -5\sqrt{2}.$$

5. $(a+b)^2$
We have
$$(a+b)^2 = \left(\sqrt{2}+\sqrt{3}+\sqrt{2}-2\sqrt{3}\right)^2$$
$$= \left(2\sqrt{2}-\sqrt{3}\right)^2$$
$$= 2^2\sqrt{2^2}-4\sqrt{6}+\sqrt{3^2}$$
$$= 4-4\sqrt{6}+3$$
$$= 7-4\sqrt{6}.$$

6. $(a-b)^2$
We have
$$(a+b)^2 = \left(\sqrt{2}+\sqrt{3}-\sqrt{2}+2\sqrt{3}\right)^2$$
$$= \left(3\sqrt{3}\right)^2$$
$$= 3^2\sqrt{3^2}$$
$$= 9 \times 3$$
$$= 27.$$

7. $(a+b)^3$
We have
$$(a+b)^3 = \left(\sqrt{2}+\sqrt{3}+\sqrt{2}-2\sqrt{3}\right)^3$$
$$= \left(2\sqrt{2}-\sqrt{3}\right)^3$$
$$= \left(2\sqrt{2}\right)^3 - 3\left(2\sqrt{2}\right)^2\left(\sqrt{3}\right) + 3\left(2\sqrt{2}\right)\left(\sqrt{3}\right)^2 - \left(\sqrt{3}\right)^3$$

6.7. Simplify Expressions in the Form of $\sqrt{a+2\sqrt{b}}$

$$= 16\sqrt{2} - 24\sqrt{3} + 18\sqrt{2} - 3\sqrt{3}$$
$$= 34\sqrt{2} - 27\sqrt{3}.$$

8. $(a-b)^3$

We have

$$(a+b)^3 = \left(\sqrt{2} + \sqrt{3} - \sqrt{2} + 2\sqrt{3}\right)^3$$
$$= \left(3\sqrt{3}\right)^3$$
$$= 3^3\sqrt{3^3}$$
$$= 27 \times 3\sqrt{3}$$
$$= 81\sqrt{3}.$$

9. $\dfrac{a^3 + a^2 b}{a+b}$

We have

$$\frac{a^3 + a^2 b}{a+b} = \frac{a^2(a+b)}{a+b}$$
$$= a^2$$
$$= \left(\sqrt{2} + \sqrt{3}\right)^2$$
$$= \sqrt{2^2} + 2\sqrt{2} \times \sqrt{3} + \sqrt{3^2}$$
$$= 2 + 2\sqrt{6} + 3$$
$$= 5 + 2\sqrt{6}.$$

10. $\dfrac{a^3 - a^2 b}{a-b}$

We have

$$\frac{a^3 - a^2 b}{a-b} = \frac{a^2(a-b)}{a-b}$$
$$= a^2$$
$$= 5 + 2\sqrt{6}.$$

Problem 13. Rationalize the following fractions:

Chapter 6. Square Roots

1. $\dfrac{1}{\sqrt{7}}$;

2. $\dfrac{1}{2\sqrt{3}}$;

3. $\dfrac{3}{\sqrt{5}}$;

4. $\dfrac{7}{2\sqrt{3}}$;

5. $\dfrac{8}{\sqrt{2}}$;

6. $\dfrac{3}{\sqrt[3]{3}}$;

7. $\dfrac{2}{\sqrt[3]{2}}$;

8. $\dfrac{5}{\sqrt[3]{4}}$;

9. $\dfrac{6}{\sqrt[3]{9}}$;

10. $\dfrac{5}{\sqrt[3]{25}}$;

Solution. Rationalize the following fractions:

1. $\dfrac{1}{\sqrt{7}}$

We have
$$\dfrac{1}{\sqrt{7}} = \dfrac{1 \times \sqrt{7}}{\sqrt{7} \times \sqrt{7}}$$
$$= \dfrac{\sqrt{7}}{\sqrt{7^2}}$$
$$= \dfrac{\sqrt{7}}{7}.$$

Therefore, $\dfrac{1}{\sqrt{7}} = \dfrac{\sqrt{7}}{7}$.

2. $\dfrac{1}{2\sqrt{3}}$

We have
$$\dfrac{1}{2\sqrt{3}} = \dfrac{1 \times \sqrt{3}}{2\sqrt{3} \times \sqrt{3}}$$
$$= \dfrac{\sqrt{3}}{2\sqrt{3^2}}$$
$$= \dfrac{\sqrt{3}}{6}.$$

Therefore, $\dfrac{1}{2\sqrt{3}} = \dfrac{\sqrt{3}}{6}$.

6.7. Simplify Expressions in the Form of $\sqrt{a+2\sqrt{b}}$

3. $\dfrac{3}{\sqrt{5}}$
   We have
   $$\dfrac{3}{\sqrt{5}} = \dfrac{3 \times \sqrt{5}}{\sqrt{5} \times \sqrt{5}}$$
   $$= \dfrac{3\sqrt{5}}{\sqrt{5^2}}$$
   $$= \dfrac{3\sqrt{5}}{5}.$$
   Therefore, $\dfrac{3}{\sqrt{5}} = \dfrac{3\sqrt{5}}{5}$.

4. $\dfrac{7}{2\sqrt{3}}$
   We have
   $$\dfrac{7}{2\sqrt{3}} = \dfrac{7 \times \sqrt{3}}{2\sqrt{3} \times \sqrt{3}}$$
   $$= \dfrac{7\sqrt{3}}{2\sqrt{3^2}}$$
   $$= \dfrac{7\sqrt{3}}{6}.$$

5. $\dfrac{8}{\sqrt{2}}$
   We have
   $$\dfrac{8}{\sqrt{2}} = \dfrac{8 \times \sqrt{2}}{\sqrt{2} \times \sqrt{2}}$$
   $$= \dfrac{8\sqrt{2}}{\sqrt{2^2}}$$
   $$= \dfrac{8\sqrt{2}}{2}$$
   $$= 4\sqrt{2}.$$

6. $\dfrac{3}{\sqrt[3]{3}}$
   We have
   $$\dfrac{3}{\sqrt[3]{3}} = \dfrac{3 \times \sqrt[3]{3^2}}{\sqrt[3]{3} \times \sqrt[3]{3^2}}$$

$$= \frac{3\sqrt[3]{9}}{\sqrt[3]{3^3}}$$
$$= \frac{3\sqrt[3]{9}}{3}$$
$$= \sqrt[3]{9}.$$

Therefore, $\frac{3}{\sqrt[3]{3}} = \sqrt[3]{9}$.

7. $\frac{2}{\sqrt[3]{2}}$

We have

$$\frac{2}{\sqrt[3]{2}} = \frac{2 \times \sqrt[3]{2^2}}{\sqrt[3]{2} \times \sqrt[3]{2^2}}$$
$$= \frac{2\sqrt[3]{4}}{\sqrt[3]{2^3}}$$
$$= \frac{2\sqrt[3]{4}}{2}$$
$$= \sqrt[3]{4}.$$

Therefore, $\frac{2}{\sqrt[3]{2}} = \sqrt[3]{4}$.

8. $\frac{5}{\sqrt[3]{4}}$

We have

$$\frac{5}{\sqrt[3]{4}} = \frac{5 \times \sqrt[3]{4^2}}{\sqrt[3]{4} \times \sqrt[3]{4^2}}$$
$$= \frac{5\sqrt[3]{16}}{\sqrt[3]{4^3}}$$
$$= \frac{5\sqrt[3]{16}}{4}.$$

Therefore, $\frac{5}{\sqrt[3]{4}} = \frac{5\sqrt[3]{16}}{4}$.

9. $\frac{6}{\sqrt[3]{9}}$

6.7. Simplify Expressions in the Form of $\sqrt{a+2\sqrt{b}}$

We have

$$\frac{6}{\sqrt[3]{9}} = \frac{6 \times \sqrt[3]{3}}{\sqrt[3]{3^2} \times \sqrt[3]{3}}$$
$$= \frac{6\sqrt[3]{3}}{\sqrt[3]{3^3}}$$
$$= \frac{6\sqrt[3]{3}}{3}$$
$$= 2\sqrt[3]{3}.$$

Therefore, $\dfrac{6}{\sqrt[3]{9}} = 2\sqrt[3]{3}$.

10. $\dfrac{5}{\sqrt[3]{25}}$
We have

$$\frac{5}{\sqrt[3]{25}} = \frac{5\sqrt[3]{5}}{\sqrt[3]{25} \times \sqrt[3]{5}}$$
$$= \frac{5\sqrt[3]{5}}{\sqrt[3]{5^3}}$$
$$= \frac{5\sqrt[3]{5}}{5}$$
$$= \sqrt[3]{5}.$$

Therefore, $\dfrac{5}{\sqrt[3]{25}} = \sqrt[3]{5}$.

Problem 14. Rationalize the following fractions:

1. $\dfrac{2}{\sqrt{3} - \sqrt{2}}$;

2. $\dfrac{4}{\sqrt{4} - \sqrt{3}}$;

3. $\dfrac{7}{2\sqrt{3} + \sqrt{2}}$;

4. $\dfrac{8}{2\sqrt{3} - 3\sqrt{2}}$;

5. $\dfrac{1}{5\sqrt{4} + 4\sqrt{5}}$;

6. $\dfrac{2}{\sqrt[3]{3} - \sqrt[3]{2}}$;

7. $\dfrac{4}{\sqrt[3]{4} - \sqrt[3]{3}}$;

8. $\dfrac{7}{2\sqrt[3]{3} + \sqrt[3]{2}}$;

Chapter 6. Square Roots

9. $\dfrac{8}{2\sqrt[3]{3} - 3\sqrt[3]{2}};$ 10. $\dfrac{1}{5\sqrt[3]{4} + 4\sqrt[3]{5}}.$

Solution. Rationalize the following fractions:

1. $\dfrac{2}{\sqrt{3} - \sqrt{2}}$
We have

$$\dfrac{2}{\sqrt{3} - \sqrt{2}} = \dfrac{2\left(\sqrt{3} + \sqrt{2}\right)}{\left(\sqrt{3} - \sqrt{2}\right)\left(\sqrt{3} + \sqrt{2}\right)}$$
$$= \dfrac{2\sqrt{3} + 2\sqrt{2}}{\sqrt{3^2} - \sqrt{2^2}}$$
$$= \dfrac{2\sqrt{3} + 2\sqrt{2}}{3 - 2}$$
$$= \dfrac{2\sqrt{3} + 2\sqrt{2}}{1}$$
$$= 2\sqrt{3} + 2\sqrt{2}.$$

Therefore, $\dfrac{2}{\sqrt{3} - \sqrt{2}} = 2\sqrt{3} + 2\sqrt{2}.$

2. $\dfrac{4}{\sqrt{4} - \sqrt{3}}$
We have

$$\dfrac{4}{\sqrt{4} - \sqrt{3}} = \dfrac{4}{2 - \sqrt{3}}$$
$$= \dfrac{4\left(2 + \sqrt{3}\right)}{\left(2 - \sqrt{3}\right)\left(2 + \sqrt{3}\right)}$$
$$= \dfrac{8 + 4\sqrt{3}}{2^2 - \sqrt{3^2}}$$
$$= \dfrac{8 + 4\sqrt{3}}{4 - 3}$$
$$= \dfrac{8 + 4\sqrt{3}}{1}$$
$$= 8 + 4\sqrt{3}.$$

Therefore, $\dfrac{4}{\sqrt{4} - \sqrt{3}} = 8 + 4\sqrt{3}.$

6.7. Simplify Expressions in the Form of $\sqrt{a + 2\sqrt{b}}$

3. $\dfrac{7}{2\sqrt{3} + \sqrt{2}}$

   We have
   $$\dfrac{7}{2\sqrt{3} + \sqrt{2}} = \dfrac{7\left(2\sqrt{3} - \sqrt{2}\right)}{\left(2\sqrt{3} + \sqrt{2}\right)\left(2\sqrt{3} - \sqrt{2}\right)}$$
   $$= \dfrac{14\sqrt{3} - 7\sqrt{2}}{\left(2\sqrt{3}\right)^2 - \sqrt{2^2}}$$
   $$= \dfrac{14\sqrt{3} - 7\sqrt{2}}{12 - 2}$$
   $$= \dfrac{14\sqrt{3} - 7\sqrt{2}}{10}.$$

   Therefore, $\dfrac{7}{2\sqrt{3} + \sqrt{2}} = \dfrac{14\sqrt{3} - 7\sqrt{2}}{10}.$

4. $\dfrac{8}{2\sqrt{3} - 3\sqrt{2}}$

   We have
   $$\dfrac{8}{2\sqrt{3} - 3\sqrt{2}} = \dfrac{8\left(2\sqrt{3} + 3\sqrt{2}\right)}{\left(2\sqrt{3} - 3\sqrt{2}\right)\left(2\sqrt{3} + 3\sqrt{2}\right)}$$
   $$= \dfrac{8\left(2\sqrt{3} + 3\sqrt{2}\right)}{\left(2\sqrt{3}\right)^2 - \left(3\sqrt{2}\right)^2}$$
   $$= \dfrac{8\left(2\sqrt{3} + 3\sqrt{2}\right)}{12 - 18}$$
   $$= \dfrac{8\left(2\sqrt{3} + 3\sqrt{2}\right)}{-6}$$
   $$= -\dfrac{4}{3}\left(2\sqrt{3} + 3\sqrt{2}\right).$$

   Therefore, $\dfrac{8}{2\sqrt{3} - 3\sqrt{2}} = -\dfrac{4}{3}\left(2\sqrt{3} + 3\sqrt{2}\right).$

5. $\dfrac{1}{5\sqrt{4} + 4\sqrt{5}}$

   We have
   $$\dfrac{1}{5\sqrt{4} + 4\sqrt{5}} = \dfrac{1}{10 + 4\sqrt{5}}$$

$$= \frac{10 - 4\sqrt{5}}{\left(10 + 4\sqrt{5}\right)\left(10 - 4\sqrt{5}\right)}$$

$$= \frac{10 - 4\sqrt{5}}{10^2 - \left(4\sqrt{5}\right)^2}$$

$$= \frac{10 - 4\sqrt{5}}{100 - 80}$$

$$= \frac{10 - 4\sqrt{5}}{20}$$

$$= \frac{10}{20} - \frac{4\sqrt{5}}{20}$$

$$= \frac{1}{2} - \frac{\sqrt{5}}{5}.$$

Therefore, $\dfrac{1}{5\sqrt{4} + 4\sqrt{5}} = \dfrac{1}{2} - \dfrac{\sqrt{5}}{5}.$

6. $\dfrac{2}{\sqrt[3]{3} - \sqrt[3]{2}}$
We have

$$\frac{2}{\sqrt[3]{3} - \sqrt[3]{2}} = \frac{2\left(\sqrt[3]{3^2} + \sqrt[3]{3} \times \sqrt[3]{2} + \sqrt[3]{2^2}\right)}{\left(\sqrt[3]{3} - \sqrt[3]{2}\right)\left(\sqrt[3]{3^2} + \sqrt[3]{3} \times \sqrt[3]{2} + \sqrt[3]{2^2}\right)}$$

$$= \frac{2\left(\sqrt[3]{9} + \sqrt[3]{6} + \sqrt[3]{4}\right)}{\sqrt[3]{3^3} - \sqrt[3]{2^3}}$$

$$= \frac{2\left(\sqrt[3]{9} + \sqrt[3]{6} + \sqrt[3]{4}\right)}{3 - 2}$$

$$= 2\left(\sqrt[3]{9} + \sqrt[3]{6} + \sqrt[3]{4}\right).$$

Therefore, $\dfrac{2}{\sqrt[3]{3} - \sqrt[3]{2}} = 2\left(\sqrt[3]{9} + \sqrt[3]{6} + \sqrt[3]{4}\right).$

7. $\dfrac{4}{\sqrt[3]{4} - \sqrt[3]{3}}$
We have

$$\frac{4}{\sqrt[3]{4} - \sqrt[3]{3}} = \frac{2\left(\sqrt[3]{4^2} + \sqrt[3]{4} \times \sqrt[3]{3} + \sqrt[3]{3^2}\right)}{\left(\sqrt[3]{4} - \sqrt[3]{3}\right)\left(\sqrt[3]{4^2} + \sqrt[3]{4} \times \sqrt[3]{3} + \sqrt[3]{3^2}\right)}$$

6.7. Simplify Expressions in the Form of $\sqrt{a + 2\sqrt{b}}$

$$= \frac{2\left(\sqrt[3]{4^2} + \sqrt[3]{4} \times \sqrt[3]{3} + \sqrt[3]{3^2}\right)}{\sqrt[3]{4^3} - \sqrt[3]{3^3}}$$

$$= \frac{2\left(\sqrt[3]{16} + \sqrt[3]{12} + \sqrt[3]{9}\right)}{4 - 3}$$

$$= 2\left(\sqrt[3]{16} + \sqrt[3]{12} + \sqrt[3]{9}\right).$$

Therefore, $\dfrac{4}{\sqrt[3]{4} - \sqrt[3]{3}} = 2\left(\sqrt[3]{16} + \sqrt[3]{12} + \sqrt[3]{9}\right).$

8. $\dfrac{7}{2\sqrt[3]{3} + \sqrt[3]{2}}$

We have

$$\frac{7}{2\sqrt[3]{3} + \sqrt[3]{2}} = \frac{7\left[\left(2\sqrt[3]{3}\right)^2 - \left(2\sqrt[3]{3}\right)\left(\sqrt[3]{2}\right) + \sqrt[3]{2^2}\right]}{\left(2\sqrt[3]{3} + \sqrt[3]{2}\right)\left[\left(2\sqrt[3]{3}\right)^2 - \left(2\sqrt[3]{3}\right)\left(\sqrt[3]{2}\right) + \sqrt[3]{2^2}\right]}$$

$$= \frac{7\left(4\sqrt[3]{9} - 2\sqrt[3]{6} + \sqrt[3]{4}\right)}{\left(2\sqrt[3]{3}\right)^3 - \sqrt[3]{2}^3}$$

$$= \frac{7\left(4\sqrt[3]{9} - 2\sqrt[3]{6} + \sqrt[3]{4}\right)}{24 - 2}$$

$$= \frac{7\left(4\sqrt[3]{9} - 2\sqrt[3]{6} + \sqrt[3]{4}\right)}{22}.$$

Therefore, $\dfrac{7}{2\sqrt[3]{3} + \sqrt[3]{2}} = \dfrac{7\left(4\sqrt[3]{9} - 2\sqrt[3]{6} + \sqrt[3]{4}\right)}{22}.$

9. $\dfrac{8}{2\sqrt[3]{3} - 3\sqrt[3]{2}}$

We have

$$\frac{8}{2\sqrt[3]{3} - 3\sqrt[3]{2}}$$

$$= \frac{8\left[\left(2\sqrt[3]{3}\right)^2 + \left(2\sqrt[3]{3}\right)\left(3\sqrt[3]{2}\right) + \left(3\sqrt[3]{2}\right)^2\right]}{\left(2\sqrt[3]{3} - 3\sqrt[3]{2}\right)\left[\left(2\sqrt[3]{3}\right)^2 - \left(2\sqrt[3]{3}\right)\left(3\sqrt[3]{2}\right) + \left(3\sqrt[3]{2}\right)^2\right]}$$

$$= \frac{8\left(4\sqrt[3]{9} + 6\sqrt[3]{6} + 9\sqrt[3]{4}\right)}{\left(2\sqrt[3]{3}\right)^3 - \left(3\sqrt[3]{2}\right)^3}$$

$$= \frac{8\left(4\sqrt[3]{9} + 6\sqrt[3]{6} + 9\sqrt[3]{4}\right)}{24 - 54}$$

$$= \frac{8\left(4\sqrt[3]{9} + 2\sqrt[3]{6} + \sqrt[3]{4}\right)}{-30}$$

$$= -\frac{4\left(4\sqrt[3]{9} + 2\sqrt[3]{6} + \sqrt[3]{4}\right)}{15}.$$

Therefore, $\dfrac{8}{2\sqrt[3]{3} - 3\sqrt[3]{2}} = -\dfrac{4\left(4\sqrt[3]{9} + 2\sqrt[3]{6} + \sqrt[3]{4}\right)}{15}.$

10. $\dfrac{1}{5\sqrt[3]{4} + 4\sqrt[3]{5}}$

We have

$$\frac{1}{5\sqrt[3]{4} + 4\sqrt[3]{5}} = \frac{\left(5\sqrt[3]{4}\right)^2 - \left(5\sqrt[3]{4}\right)\left(4\sqrt[3]{5}\right) + \left(4\sqrt[3]{5}\right)^2}{\left(5\sqrt[3]{4} + 4\sqrt[3]{5}\right)\left[\left(5\sqrt[3]{4}\right)^2 - \left(5\sqrt[3]{4}\right)\left(4\sqrt[3]{5}\right) + \left(4\sqrt[3]{5}\right)^2\right]}$$

$$= \frac{25\sqrt[3]{16} - 20\sqrt[3]{20} + 16\sqrt[3]{25}}{\left(5\sqrt[3]{4}\right)^3 - \left(4\sqrt[3]{5}\right)^3}$$

$$= \frac{25\sqrt[3]{16} - 20\sqrt[3]{20} + 16\sqrt[3]{25}}{500 - 320}$$

$$= \frac{25\sqrt[3]{16} - 20\sqrt[3]{20} + 16\sqrt[3]{25}}{180}.$$

Therefore, $\dfrac{1}{5\sqrt[3]{4} + 4\sqrt[3]{5}} = \dfrac{25\sqrt[3]{16} - 20\sqrt[3]{20} + 16\sqrt[3]{25}}{180}.$

Chapter 7

Linear Equations in One Variable

Solving equations is the heart of learning Algebra. Therefore, equations are the most important part that learners have to understand about. There are many types of equations. In this chapter, we will introduce readers about Linear Equations in One Variable. There are some basics that the readers have to know before starting the concepts of how to solve Linear Equations in One Variable.

7.1 Defintions

1. **Equations**
   An equation is a mathematical statement of an equality that contains one or more variables.

   > **Example 41**
   > $3x - 2 = 0$, $3x + 1 = 2x$, $3y + 5 = 2y - 1$, $x + y = z$
   > are called equations.

2. **Linear Equations in One Variable**
   A linear equation in one variable is an equation that can be written in the form $ax + b = c$, where a, b and c are real numbers and $a \neq 0$. Notice that linear equations in one variable can be also called first-degree equations in one variable.

Chapter 7. Linear Equations in One Variable

> **Example 42**
> $2x - 4 = 2$, $4y = 3y - 2$, $4x + 4 = 0$ are called linear equations in one variable.

3. **Solution**
A solution of an equation is a value that satisfies the equation. That is, when we replace x by that value, we obtain the right-hand side equals the left-hand side. Note that a solution of an equation can be called its root. That is, solutions and roots have the same meaning.

> **Example 43**
> 1 is the solution of the equation $3x + 2 = 5$.
> Observe that when we replace x by 1 we obtain $3x + 2 = 3(1) + 2 = 5$, true.

> **Example 44**
> Prove that $x = 2$ is a root of the equation
> $$x^3 - 8x + 8 = 0.$$

Solution. Substitute x by 2, it follows that

$$x^3 - 8x + 8 = 2^3 - 8(2) + 8 = 8 - 16 + 8 = 0, \text{true}.$$

7.2 Operation Properties of Equality

For all real numbers a, b and c, the following properties hold:

1. If $a = b$, then $a + c = b + c$;

2. If $a = b$, then $a - c = b - c$;

3. If $a = b$, then $ac = bc$ for all $c \neq 0$;

4. if $a = b$, then $\dfrac{a}{c} = \dfrac{b}{c}$ for all $c \neq 0$.

In other words, if two expressions are equal to each other and we add, subtract, divide or multiply the exact same thing to both sides, the two sides will remain equal.

Since we have known that subtraction and division are the inverse of addition and multiplication respectively, to move a number from one side to another side in an equation, we have to use this features. That is if we have a number that is being added and we want to move it to another side, we would subtract it from both sides of the equation.

7.3 How to Solve It

To solve linear equations in one variable, we have to separate variable and constants by using inverse operations. See the following examples.

Example 45

Solve the following equations:

1. $5x - 2 = 3$;
2. $5(x - 1) + 2(x - 3) = 2x - 3$;
3. $2x - 3 + 5x - 7 = 6x + 8$;
4. $2(x - 1) + 3(x - 2) + 4(x - 5) = 3$;
5. $(x+1)(x+2) + (x+2)(x+3) = (x+3)(x+4) + (x+4)(x+5)$.

Solution. Solve the following equations:

1. $5x - 2 = 3$

    We have

$5x - 2 = 3$

$5x - 2 + 2 = 3 + 2$ add 2 to both sides to omit -2

$5x = 5$

$\dfrac{5x}{5} = \dfrac{5}{5}$ divide both sides by 5 to omit the coefficient of x

$x = 1.$

Consequently, $x = 1$ is the solution.

2. $5(x-1) + 2(x-3) = 2x - 3$
We have
$$5(x-1) + 2(x-3) = 2x - 3$$
$$5x - 5 + 2x - 6 = 2x - 3$$
$$7x - 11 = 2x - 3$$
$$7x - 2x = -3 + 11$$
$$5x = 8$$
$$x = \frac{8}{5}.$$

Thus, $x = \dfrac{8}{5}$ is the solution.

3. $2x - 3 + 5x - 7 = 6x + 8$
We have
$$2x - 3 + 5x - 7 = 6x + 8$$
$$7x - 10 = 6x + 8$$
$$7x - 6x = 8 + 10$$
$$x = 18.$$

Thus, $x = 18$ is the solution.

4. $2(x-1) + 3(x-2) + 4(x-5) = 3$
We have
$$2(x-1) + 3(x-2) + 4(x-5) = 3$$
$$2x - 2 + 3x - 6 + 5x - 20 = 3$$
$$10x - 28 = 3$$
$$10x = 3 + 28$$
$$10x = 31$$
$$x = \frac{31}{10}.$$

Therefore, $x = \dfrac{31}{10}$ is the solution.

7.3. How to Solve It

5. $(x+1)(x+2)+(x+2)(x+3) = (x+3)(x+4)+(x+4)(x+5)$

We have

$$(x+1)(x+2) + (x+2)(x+3)$$
$$= (x+3)(x+4) + (x+4)(x+5)$$
$$x^2 + 2x + x + 2 + x^2 + 3x + 2x + 6$$
$$= x^2 + 4x + 3x + 12 + x^2 + 4x + 5x + 20$$
$$2x^2 + 8x + 8 = 2x^2 + 16x + 32$$
$$2x^2 + 8x - 2x^2 - 16x = 32 - 8$$
$$-8x = 24$$
$$x = \frac{24}{-8} = -3.$$

Consequently, $x = -3$ is the solution.

To be convenient in solving equations, we should know that when we change a term from one side to another side we have to change its sign. Namely, we have to obey the following rules:

- Addition change to subtraction;
- Subtraction change to Addition;
- Multiplication change to division;
- Division change to multiplication.

Example 46

Solve the following equations:

1. $\dfrac{x+1}{2} - 3 = \dfrac{1}{2}$;

2. $\dfrac{x-3}{2} - \dfrac{x}{3} = \dfrac{1}{2} + x$;

3. $x + 2(x-1) + \dfrac{x}{3} = 4x - 4$;

4. $\dfrac{x-1}{2017} + \dfrac{x-2}{2016} + \dfrac{x-3}{2015} + \dfrac{x-4}{2014} = 4.$

Solution. Solve the following equations:

Chapter 7. Linear Equations in One Variable

1. $\dfrac{x+1}{2} - 3 = \dfrac{1}{2}$

   We have
   $$\dfrac{x+1}{2} - 3 = \dfrac{1}{2}$$
   $$\dfrac{x+1-6}{2} = \dfrac{1}{2}$$
   $$x - 5 = 1$$
   $$x = 1 + 5 = 6.$$

2. $\dfrac{x-3}{2} - \dfrac{x}{3} = \dfrac{1}{2} + x$

   We have
   $$\dfrac{x-3}{2} - \dfrac{x}{3} = \dfrac{1}{2} + x$$
   $$\dfrac{3(x-3) - 2x}{6} = \dfrac{3 + 6x}{6}$$
   $$3x - 9 - 2x = 3 + 6x$$
   $$x - 9 = 3 + 6x$$
   $$x - 6x = 3 + 9$$
   $$-5x = 12$$
   $$x = -\dfrac{12}{5}.$$

3. $x + 2(x-1) + \dfrac{x}{3} = 4x - 4$

   We have
   $$x + 2(x-1) + \dfrac{x}{3} = 4x - 4$$
   $$x + 2x - 2 + \dfrac{x}{3} = 4x - 4$$
   $$3x - 2 + \dfrac{x}{3} = 4x - 4$$
   $$\dfrac{9x - 6 + x}{3} = \dfrac{12x - 12}{3}$$
   $$10x - 6 = 12x - 12$$
   $$10x - 12x = -12 + 6$$
   $$-2x = -6$$

7.3. How to Solve It

$$x = \frac{-6}{-2} = 3.$$

4. $\dfrac{x-1}{2017} + \dfrac{x-2}{2016} + \dfrac{x-3}{2015} + \dfrac{x-4}{2014} = 4.$
We have
$$\frac{x-1}{2017} + \frac{x-2}{2016} + \frac{x-3}{2015} + \frac{x-4}{2014} = 4$$
$$\frac{x-1}{2017} - 1 + \frac{x-2}{2016} - 1 + \frac{x-3}{2015} - 1 + \frac{x-4}{2014} - 1 = 0$$
$$\frac{x-2018}{2017} + \frac{x-2018}{2016} + \frac{x-2018}{2015} + \frac{x-2018}{2014} = 0$$
$$(x-2018)\left(\frac{1}{2017} + \frac{1}{2016} + \frac{1}{2015} + \frac{1}{2014}\right) = 0$$
$$x - 2018 = 0$$
$$x = 2018.$$

> **Practice 3**
>
> Solve the following equations:
>
> 1. $2x - 5 = 7x - 2$;
> 2. $-3x - 6 = -(x-1) - x + 2$;
> 3. $4\left(x - \dfrac{1}{2}\right) + 100\left(\dfrac{x}{50} - 1\right) = 90 + x$;
> 4. $2(x-1) + 5(x-7) = 3(x-3)$;
> 5. $x[2(x-1) + 3(x-2)] = (5x-1)(x+2)$;
> 6. $(x-1)^2 + (x-2)^2 + (x-3)^2 = 3x^2 - 7x$;
> 7. $\dfrac{x-2}{3} + \dfrac{x+1}{5} - 1 = \dfrac{1-x}{7}$;
> 8. $x - \dfrac{1-x}{2} = 3(x-1) + 5x$;
> 9. $\dfrac{x-b}{a} + \dfrac{x-a}{b} = (a+b+x) - (a+b-x)$;
> 10. $\dfrac{1}{5}\left[2(1-x) - \dfrac{1}{7}(2-x)\right] = \dfrac{5-x}{3}$.

Chapter 7. Linear Equations in One Variable

Exercises

Problem 1. Solve the following equations:

1. $3(x-2) - 2(x-1) = 3$;

2. $2x - 5(x-2) = -3x + 2$;

3. $2x - (x-1) + 3 = \dfrac{1}{2}(2x - 8)$;

4. $3(x-1) - 2(x-2) + 4(x-5) = 4(7-x)$;

5. $2(x-3) + 6(x-3) = 4x - 5$.

Problem 2. Solve the following equations:

1. $\dfrac{x}{3} + \dfrac{x}{2} + \dfrac{x}{101} + \dfrac{x}{102} = 0$;

2. $\dfrac{x-1}{2} + \dfrac{x}{3} - \dfrac{x-2}{6} = 3 + x$;

3. $\dfrac{1-x}{3} - 2(x-2) + 3(x-4) = \dfrac{x}{6}$;

4. $\dfrac{x-3}{5} - 3x = \dfrac{x}{2} - 3$.

Problem 3. Find the value of m such that the following equation has no roots
$$(m-1)x + 4m = 4x - 3.$$

Problem 4. Find the value of m such that the following equation has only a solution

$$(x-2)m + 3x - 3 = 2(x-4).$$

Problem 5. Find the value of m such that the equation

$$4(m-2)x + \frac{1}{3}[9(x-2)] = 3x - 6$$

has infinitely many solutions.

Solutions

Problem 1. Solve the following equations:

1. $3(x-2) - 2(x-1) = 3$;

2. $2x - 5(x-2) = -3x + 2$;

3. $2x - (x-1) + 3 = \dfrac{1}{2}(2x - 8)$;

4. $3(x-1) - 2(x-2) + 4(x-5) = 4(7-x)$;

5. $2(x-3) + 6(x-3) = 4x - 5$.

Solution. Solve the following equations:

1. $3(x-2) - 2(x-1) = 3$
We have
$$3(x-2) - 2(x-1) = 3$$
$$3x - 6 - 2x + 2 = 3$$
$$x - 4 = 3$$
$$x = 3 + 4$$
$$x = 7.$$

2. $2x - 5(x-2) = -3x + 2$
We have
$$2x - 5(x-2) = -3x + 2$$

$$2x - 5x + 10 = -3x + 2$$
$$-3x + 10 = -3x + 2$$
$$-3x + 3x = 2 - 10$$
$$0 = -8, \quad \text{not true.}$$

Thus, the given equation has no roots.

3. $2x - (x - 1) + 3 = \dfrac{1}{2}(2x + 8)$

We have
$$2x - (x - 1) + 3 = \dfrac{1}{2}(2x + 8)$$
$$2x - x + 1 + 3 = x + 4$$
$$x + 4 = x + 4$$
$$0 = 0.$$

It is true for all real numbers x.

Thus, the given equation has infinitely many solutions.

4. $3(x - 1) - 2(x - 2) + 4(x - 5) = 4(7 - x)$

We have
$$3(x - 1) - 2(x - 2) + 4(x - 5) = 4(7 - x)$$
$$3x - 3 - 2x + 4 + 4x - 20 = 28 - 4x$$
$$5x - 19 = 28 - 4x$$
$$5x + 4x = 28 + 19$$
$$9x = 47$$
$$x = \dfrac{47}{9}.$$

5. $2(x - 3) + 6(x - 3) = 4x - 5$

We have
$$2(x - 3) + 6(x - 3) = 4x - 5$$
$$2x - 6 + 6x - 18 = 4x - 5$$
$$8x - 24 = 4x - 5$$
$$8x - 4x = 24 - 5$$
$$4x = 19$$
$$x = \dfrac{19}{4}.$$

7.3. How to Solve It

Problem 2. Solve the following equations:

1. $\dfrac{x}{3} + \dfrac{x}{2} + \dfrac{x}{101} + \dfrac{x}{102} = 0;$

2. $\dfrac{x-1}{2} + \dfrac{x}{3} - \dfrac{x-2}{6} = 3 + x;$

3. $\dfrac{1-x}{3} - 2(x-2) + 3(x-4) = \dfrac{x}{6};$

4. $\dfrac{x-3}{5} - 3x = \dfrac{x}{2} - 3.$

Solution. Solve the following equations:

1. $\dfrac{x}{3} + \dfrac{x}{2} + \dfrac{x}{101} + \dfrac{x}{102} = 0$
We have
$$\dfrac{x}{3} + \dfrac{x}{2} + \dfrac{x}{101} + \dfrac{x}{102} = 0$$
$$x\left(\dfrac{1}{3} + \dfrac{1}{2} + \dfrac{1}{101} + \dfrac{1}{102}\right) = 0$$
$$x = 0.$$

2. $\dfrac{x-1}{2} + \dfrac{x}{3} - \dfrac{x-2}{6} = 3 + x$
We have
$$\dfrac{x-1}{2} + \dfrac{x}{3} - \dfrac{x-2}{6} = 3 + x$$
$$3(x-1) + 2x - (x-2) = 6(3+x)$$
$$3x - 3 + 2x - x + 2 = 18 + 6x$$
$$4x - 1 = 18 + 6x$$
$$4x - 6x = 18 + 1$$
$$-2x = 19$$
$$x = -\dfrac{19}{2}.$$

3. $\dfrac{1-x}{3} - 2(x-2) + 3(x-4) = \dfrac{x}{6}$
We have
$$\dfrac{1-x}{3} - 2(x-2) + 3(x-4) = \dfrac{x}{6}$$

Chapter 7. Linear Equations in One Variable

$$2(1-x) - 12(x-2) + 18(x-4) = x$$
$$2 - 2x - 12x + 24 + 18x - 72 = x$$
$$4x - 46 = x$$
$$4x - x = 46$$
$$3x = 46$$
$$x = \frac{46}{3}.$$

4. $\dfrac{x-3}{5} - 3x = \dfrac{x}{2} - 3$

We have

$$\frac{x-3}{5} - 3x = \frac{x}{2} - 3$$
$$2(x-3) - 30x = 5x - 30$$
$$2x - 6 - 30x = 5x - 30$$
$$-28x - 6 = 5x - 30$$
$$-28x - 5x = -30 + 6$$
$$-33x = -24$$
$$x = \frac{-24}{-33} = \frac{8}{11}.$$

Problem 3. Find the value of m such that the following equation has no roots
$$(m-1)x + 4m = 4x - 3.$$

Solution. Find the value of m.
We have

$$(x-2)m + 3x - 3 = 2(x-4)$$
$$mx - 2m + 3x - 3 = 2x - 8$$
$$mx + 3x - 2x = -8 + 2m + 3$$
$$mx + x = 2m - 5$$
$$(m+1)x = 2m - 5.$$

The equation has no roots if and only if $\begin{cases} m+1 = 0 \\ 2m - 5 \neq 0 \end{cases}$ or $\begin{cases} m = -1 \\ m \neq \dfrac{5}{2} \end{cases}$.

Therefore, $m = -1$.

7.3. How to Solve It

Problem 4. Find the value of m such that the following equation has only a solution
$$(x-2)m + 3x - 3 = 2(x-4).$$
Solution. Find the value of m.
We have
$$(m-1)x + 4m = 4x - 3$$
$$mx - x + 4m = 4x - 3$$
$$mx - x - 4x = -4m - 3$$
$$mx - 5x = -4m - 3$$
$$(m-5)x = -4m - 3.$$

The equation has only a solution if and only if
$$\begin{cases} m - 5 \neq 0 \\ -4m - 3 \neq 0 \end{cases}$$
or
$$\begin{cases} m \neq 5 \\ m \neq -\dfrac{3}{4} \end{cases}.$$

Therefore, the given equation has only a solution when $\begin{cases} m \neq 5 \\ m \neq -\dfrac{3}{4} \end{cases}$.

Its solution is $x = -\dfrac{4m+3}{m-5}$.

Problem 5. Find the value of m such that the equation
$$4(m-2)x + \frac{1}{3}[9(x-2)] = 3x - 6$$
has infinitely many solutions.

Solution. Find the value of m.
We have
$$4(m-2)x + \frac{1}{3}[9(x-2)] = 3x - 6$$
$$4(m-2)x + 3(x-2) = 3x - 6$$
$$4(m-2)x + 3x - 6 = 3x - 6$$
$$4(m-2)x = 0.$$

Therefore, the given equation has infinitely many solutions if and only if $m - 2 = 0$. It holds when $m = 2$.

Chapter 7. Linear Equations in One Variable

Chapter 8

Linear Inequalities With One Variable

Linear inequalities in one variable are not just theoretical concepts; they have significant mathematical and practical importance. In this chapter, we will introduce readers to the definition of linear inequalities in one variable and how to solve it.

8.1 Defintion

A linear inequality with one variable is the inequality of the form $ax + b > 0, ax + b < 0, ax + b \geq 0$ or $ax + b \leq 0$, where a and b are real numbers with $a \neq 0$.

Example 47

$2x - 1 > 0, -3x - 5 \leq 0, 7x - 5 \geq 0$ and $8x - 5 \geq 3x - 1$ are linear inequalities with one variable.

8.2 Properties of Inequality

For all real numbers a, b and c, we obtain the following properties:

1. If $a > b$, then $a + c > b + c$;
2. If $a > b$, then $a - c > b - c$;

3. If $a > b$ and $c > 0$, we obtain $ac > bc$ and $\dfrac{a}{c} > \dfrac{b}{c}$;

4. If $a > b$ and $c < 0$, we obtain $ac < bc$ and $\dfrac{a}{c} < \dfrac{b}{c}$.

8.3 How To Solve It

It is not hard to solve linear inequalities with one variable if we know how to solve linear equations with one variable (See Chapter V). The given examples will explain readers how to linear equations with one variable.

Example 48

Solve the following inequalities:

1. $3x - 5 \geq 2x - 6$;
2. $-4x + 5 < 3 - 2x$;
3. $4(x-1) - 2(5+x) \geq 3x - 5$;
4. $\dfrac{1}{3}x + 2(x-1) > 3 + 5(3-x)$;
5. $2(x+1) - 3(x+2) \leq 3(x-2) + 5x$.

Solution. Solve the following inequalities:

1. $3x - 5 \geq 2x - 6$
   We have
   $$3x - 5 \geq 2x - 6$$
   $$3x - 2x \geq -6 + 5$$
   $$x \geq -1.$$

2. $-4x + 5 < 3 - 2x$
   We have
   $$-4x + 5 < 3 - 2x$$
   $$-4x + 2x < 3 - 5$$
   $$-2x < -2$$
   $$x > \dfrac{-2}{-2} = 1.$$

8.3. How To Solve It

3. $4(x-1) - 2(5+x) \geq 3x - 5$
We have
$$4(x-1) - 2(5+x) \geq 3x - 5$$
$$4x - 4 - 10 - 2x \geq 3x - 5$$
$$2x - 14 \geq 3x - 5$$
$$2x - 3x \geq -5 + 14$$
$$-x \geq 9$$
$$x \leq -9.$$

4. $\frac{1}{3}x + 2(x-1) > 3 + 5(3-x)$
We have
$$\frac{1}{3}x + 2(x-1) > 3 + 5(3-x)$$
$$\frac{1}{3}x + 2x - 2 > 3 + 15 - 5x$$
$$x + 6x - 6 > 9 + 45 - 15x$$
$$7x - 6 > 54 - 15x$$
$$7x + 15x > 54 + 6$$
$$22x > 60$$
$$x > \frac{60}{22} = \frac{30}{11}.$$

5. $2(x+1) - 3(x+2) \leq 3(x-2) + 5x$
We have
$$2(x+1) - 3(x+2) \leq 3(x-2) + 5x$$
$$2x + 2 - 3x - 6 \leq 3x - 6 + 5x$$
$$-x - 4 \leq 8x - 6$$
$$-x - 8x \leq -6 + 4$$
$$-9x \leq -2$$
$$x \geq \frac{-2}{-9} = \frac{2}{9}.$$

Chapter 8. Linear Inequalities With One Variable

> **Example 49**
>
> Solve the following inequalities:
>
> 1. $\dfrac{3x}{2} - 2(x-1) > 3 + \dfrac{x}{2}$;
>
> 2. $\dfrac{3(x-1)}{7} - 1 \leq x + \dfrac{3}{2}$;
>
> 3. $\dfrac{5(x-1)}{6} - 3x \geq 7x - \dfrac{1}{2}$;
>
> 4. $7(3-x) + 6x > 3(x-1) + \dfrac{x}{2}$;
>
> 5. $3(4-x) + 5x \leq 2(2-x) + 6x$.

Solution. Solve the following inequalities:

1. $\dfrac{3x}{2} - 2(x-1) > 3 + \dfrac{x}{2}$
We have
$$\dfrac{3x}{2} - 2(x-1) > 3 + \dfrac{x}{2}$$
$$\dfrac{3x - 4(x-1)}{2} > \dfrac{6+x}{2}$$
$$3x - 4x + 4 > 6 + x$$
$$-x + 4 > 6 + x$$
$$-x - x > 6 - 4$$
$$-2x > 2$$
$$x < \dfrac{2}{-2} = -1.$$

2. $\dfrac{3(x-1)}{7} - 1 \leq x + \dfrac{3}{2}$
We have
$$\dfrac{3(x-1)}{7} - 1 \leq x + \dfrac{3}{2}$$
$$\dfrac{6(x-1)}{14} - \dfrac{14}{14} \leq \dfrac{14x + 21}{14}$$
$$6x - 6 - 14 \leq 14x + 21$$

8.3. How To Solve It

$$6x - 20 \leq 14x + 21$$
$$6x - 14x \leq 21 + 20$$
$$-8x \leq 41$$
$$x \geq -\frac{41}{8}.$$

3. $\dfrac{5(x-1)}{6} - 3x \geq 7x - \dfrac{1}{2}$

We have
$$\frac{5(x-1)}{6} - 3x \geq 7x - \frac{1}{2}$$
$$\frac{5(x-1) - 18x}{6} \geq \frac{42x - 3}{6}$$
$$5(x-1) - 18x \geq 42x - 3$$
$$5x - 5 - 18x \geq 42x - 3$$
$$-13x - 42x \geq -3 + 5$$
$$-55x \geq 2$$
$$x \leq -\frac{2}{55}$$

4. $7(3-x) + 6x > 3(x-1) + \dfrac{x}{2}$

We have
$$7(3-x) + 6x > 3(x-1) + \frac{x}{2}$$
$$21 - 7x + 6x > 3x - 3 + \frac{x}{2}$$
$$21 - x > 3x - 3 + \frac{x}{2}$$
$$\frac{42 - 2x}{2} > \frac{6x - 6 + x}{2}$$
$$42 - 2x > 7x - 6$$
$$-2x - 7x > -6 - 42$$
$$-9x > -48$$
$$x < \frac{-48}{-9} = \frac{48}{9}.$$

5. $3(4-x) + 5x \leq 2(2-x) + 6x$

We have
$$3(4-x) + 5x \leq 2(2-x) + 6x$$

Chapter 8. Linear Inequalities With One Variable

$$12 - 3x + 5x \leq 4 - 2x + 6x$$
$$12 + 2x \leq 4 + 4x$$
$$2x - 4x \leq 4 - 12$$
$$-2x \leq -8$$
$$x \geq \frac{-8}{-2} = 4.$$

Chapter 9

Quadratic Equations

A quadratic equation is a type of polynomial equation that plays a central role in algebra and mathematical modeling. The defining feature of a quadratic equation is the x^2 term, which creates a parabolic curve when graphed, making these equations particularly useful in describing various real-world phenomena like projectile motion, optimization problems, and physics scenarios. Whether in ancient civilizations or modern applications, quadratic equations have been essential in advancing our understanding of both mathematical theory and practical problem-solving. Their rich history and versatility continue to make them an important topic of study for students and professionals alike.

9.1 Definition

A quadratic equation is an equation that can be written in the form $ax^2 + bx + c = 0$, where a, b and c are real numbers with $a \neq 0$. a, b and c are called the coefficients of the equation.

Example 50

$x^2 - x + 1 = 0, x^2 - 4x + 2 = 0, 2x^2 - 3x - 7 = 0, \ldots$ are called the quadratic equations.

Remark 8.
- a is called quadratic coefficient
- b is called linear coefficient

- and c is called constant term or free term.

9.2 How To Solve It

There are many methods in solving quadratic equations. In this chapter, we will introduce readers about some important methods.

9.2.1 Product Equals Zero

> **Theorem 3**
> Suppose that A and B are real numbers. It follows that $AB = 0$ if and only if $A = 0$ or $B = 0$.

Example 51

Solve the following equations:

1. $(x-1)(x-2) = 0$;
2. $(x-4)(-x+3) = 0$;
3. $x(x-2) + 3(x-2) = 0$;
4. $x^2 - 5x + 6 = 0$;
5. $x^2 - 4x + 3 = 0$.

Solution. Solve the following equations:

1. $(x-1)(x-2) = 0$

    We have $(x-1)(x-2) = 0$.
    Using the above theorem, we obtain $\begin{bmatrix} x-1 = 0 \\ x-2 = 0 \end{bmatrix}$ or $\begin{bmatrix} x = 1 \\ x = 2 \end{bmatrix}$.

2. $(x-4)(-x+3) = 0$

    We have $(x-4)(-x+3) = 0$.
    It follows that $\begin{bmatrix} x-4 = 0 \\ -x+3 = 0 \end{bmatrix}$ or $\begin{bmatrix} x = 4 \\ x = 3 \end{bmatrix}$.

9.2. How To Solve It

3. $x(x-2) + 3(x-2) = 0$

    We have
    $$x(x-2) + 3(x-2) = 0$$
    $$(x-2)(x+3) = 0$$

    It implies that $\left[\begin{array}{l} x+3=0 \\ x-2=0 \end{array}\right.$ or $\left[\begin{array}{l} x=-3 \\ x=2 \end{array}\right.$.

4. $x^2 - 5x + 6 = 0$

    We have
    $$x^2 - 5x + 6 = 0$$
    $$x^2 - 2x - 3x + 6 = 0$$
    $$x(x-2) - 3(x-2) = 0$$
    $$(x-2)(x-3) = 0$$

    Consequently, $\left[\begin{array}{l} x-2=0 \\ x-3=0 \end{array}\right.$ or $\left[\begin{array}{l} x=2 \\ x=3 \end{array}\right.$.

5. $x^2 - 4x + 3 = 0$

    We have
    $$x^2 - 4x + 3 = 0$$
    $$x^2 - x - 3x + 3 = 0$$
    $$x(x-1) - 3(x-1) = 0$$
    $$(x-1)(x-3) = 0$$

    Therefore, $\left[\begin{array}{l} x-1=0 \\ x-3=0 \end{array}\right.$ or $\left[\begin{array}{l} x=1 \\ x=3 \end{array}\right.$.

> **Practice 4**
>
> Solve the following equations:
>
> 1. $x(x-7) + 2(x-7) = 0$;
> 2. $x(x-9) + 8 = 0$;
> 3. $x^2 - 6x + 8 = 0$;

4. $x^2 - 7x + 10 = 0$;

5. $(x+1)(x+2) + (x+2)(x+3) = 0$.

9.2.2 Quadratic In Form: $x^2 = a$, where a is a positive real number

From square root's definition, we have $x^2 = a$ if and only if $x = \pm\sqrt{a}$, where a is a positive real number.

Example 52

Solve the following equations:

1. $x^2 = 9$;
2. $x^2 - 4 = 0$;
3. $x^2 = 3$;
4. $x^2 - 2 = 0$.

Solution. Solve the following equations:

1. $x^2 = 9$
   We have $x^2 = 9$. Then $x = \pm\sqrt{9} = \pm 3$.

2. $x^2 - 4 = 0$
   We have $x^2 - 4 = 0$ or $x^2 = 4$. It follows that $x = \pm\sqrt{4} = \pm 2$.

3. $x^2 = 3$
   We have $x^2 = 3$. It implies that $x = \pm\sqrt{3}$.

4. $x^2 - 2 = 0$
   We have $x^2 - 2 = 0$ or $x^2 = 2$. Then $x = \pm\sqrt{2}$.

Practice 5

Solve the following equations:

1. $2x^2 - 8 = 0$;
2. $4x^2 - 16 = 0$;
3. $4(x^2 - x) + 6 = -4(x - 2)$;
4. $(x+1)(x-1) = (2x-3)(2x+3)$.

9.2. How To Solve It

9.2.3 Completing Square

Completing Square is another method in solving quadratic equation. To see how this method is used, we observe the following examples.

Example 53

Solve the following equation:
$$x^2 + 6x - 8 = 0.$$

Solution. We cannot factor the right-hand side of the given equation. Moreover, the given equation cannot be solved directly by using the definition of square root. To solve the given equation, we complete the right-hand side of the equation to a square of a binomial.

That is,

$$x^2 + 6x - 8 = 0$$
$$x^2 + 6x = 8$$
$$x^2 + 2(x)(3) + 3^2 = 8 + 3^2$$
$$(x+3)^2 = 17.$$

It follows that $x + 3 = \pm\sqrt{17}$. Then $x = -3 \pm \sqrt{17}$. Thus, $x = -3 \pm \sqrt{17}$.

Example 54

Solve the following equations:

1. $x^2 + 4x + 1 = 0$;
2. $x^2 - 2x - 5 = 0$;
3. $x^2 - 8x - 7 = 0$;
4. $x^2 + 5x - 9 = 0$;
5. $x^2 - 3x - 11 = 0$.

Solution. Solve the following equations:

Chapter 9. Quadratic Equations

1. $x^2 + 4x + 1 = 0$
We have
$$x^2 + 4x + 1 = 0$$
$$x^2 + 4x = -1$$
$$x^2 + 2(x)(2) + 2^2 = 2^2 - 1$$
$$(x+2)^2 = 3.$$

It implies that $x + 2 = \pm\sqrt{3}$. Then $x = -2 \pm \sqrt{3}$.

2. $x^2 - 2x - 5 = 0$
We have
$$x^2 - 2x - 5 = 0$$
$$x^2 - 2x = 5$$
$$x^2 - 2(x)(1) + 1^2 = 5 + 1^2$$
$$(x-1)^2 = 6.$$

Thus, $x - 1 = \pm\sqrt{6}$. Then $x = 1 \pm \sqrt{6}$.

3. $x^2 - 8x - 7 = 0$
We have
$$x^2 - 8x - 7 = 0$$
$$x^2 - 8x = 7$$
$$x^2 - 2(x)(4) + 4^2 = 7 + 4^2$$
$$(x-4)^2 = 23.$$

We obtain $x - 4 = \pm\sqrt{23}$. Then $x = 4 \pm \sqrt{23}$.

4. $x^2 + 5x - 9 = 0$
We have
$$x^2 + 5x - 9 = 0$$
$$x^2 + 5x = 9$$
$$x^2 + 2(x)\left(\frac{5}{2}\right) + \left(\frac{5}{2}\right)^2 = 9 + \left(\frac{5}{2}\right)^2$$
$$\left(x + \frac{5}{2}\right)^2 = 9 + \frac{25}{4}$$

9.2. How To Solve It

$$\left(x+\frac{5}{2}\right)^2 = \frac{61}{4}.$$

We obtain $x + \dfrac{5}{2} = \pm\sqrt{\dfrac{61}{4}} = \pm\dfrac{\sqrt{61}}{2}.$

Then $x = -\dfrac{5}{2} \pm \dfrac{\sqrt{61}}{2}.$

5. $x^2 - 3x - 11 = 0$
   We have

$$x^2 - 3x - 11 = 0$$
$$x^2 - 3x = 11$$
$$x^2 - 2(x)\left(\frac{3}{2}\right) + \left(\frac{3}{2}\right)^2 = 11 + \left(\frac{3}{2}\right)^2$$
$$\left(x - \frac{3}{2}\right)^2 = 11 + \frac{9}{4}$$
$$\left(x - \frac{3}{2}\right)^2 = \frac{53}{4}.$$

It implies that $x - \dfrac{3}{2} = \pm\sqrt{\dfrac{53}{4}}.$

Consequently, $x = \dfrac{3}{2} \pm \dfrac{\sqrt{53}}{2}.$

9.2.4 Solve Quadratic Equations By Discriminant

Solve Quadratic Equations by Discriminant is a really important method for solving Quadratic equation. This method can help us to solve all quadratic equations. The formula in this method is generated from the completing square method.

> Theorem 4
>
> Given a quadratic equation $ax^2 + bx + c = 0$ where a, b and c are real numbers and $a \neq 0$. We define $\Delta = b^2 - 4ac$. Δ is called the discriminant of the quadratic equation.
>
> - If $\Delta > 0$, the given equation has two distinct roots.

They are
$$x_1 = \frac{-b + \sqrt{\Delta}}{2a}$$
and
$$x_2 = \frac{-b - \sqrt{\Delta}}{2a}.$$

- If $\Delta = 0$, the given equation has two roots with the same values. That is,
$$x_1 = x_2 = -\frac{b}{2a}.$$

- If $\Delta < 0$, the given equation has no real roots.

Proof. We have
$$ax^2 + bx + c = 0$$
$$x^2 + \frac{b}{a}x + \frac{c}{a} = 0$$
$$x^2 + \frac{b}{a}x = -\frac{c}{a}$$
$$x^2 + 2(x)\left(\frac{b}{2a}\right) + \left(\frac{b}{2a}\right)^2 = \left(\frac{b}{2a}\right)^2 - \frac{c}{a}$$
$$\left(x + \frac{b}{2a}\right)^2 = \frac{b^2}{4a^2} - \frac{c}{a}$$
$$\left(x + \frac{b}{2a}\right)^2 = \frac{b^2 - 4ac}{4a^2}$$
$$\left(x + \frac{b}{2a}\right)^2 = \frac{\Delta}{4a^2}.$$

- If $\Delta < 0$, we obtain $\left(x + \frac{b}{2a}\right)^2 < 0$, a contradiction. Consequently, the given equation has no real roots.

- If $\Delta = 0$, we obtain $\left(x + \frac{b}{2a}\right)^2 = 0$. Hence, $x_1 = x_2 = -\frac{b}{2a}$.

9.2. How To Solve It

- If $\Delta > 0$, we obtain $x + \dfrac{b}{2a} = \pm\sqrt{\dfrac{\Delta}{4a^2}} = \pm\dfrac{\sqrt{\Delta}}{2a}$. Then $x = \dfrac{-b \pm \sqrt{\Delta}}{2a}$. Therefore, $x_1 = \dfrac{-b + \sqrt{\Delta}}{2a}$ and $x_2 = \dfrac{-b - \sqrt{\Delta}}{2a}$. □

> **Example 55**
> Solve the following equations:
> 1. $x^2 - 5x + 3 = 0$;
> 2. $2x^2 - 3x - 4 = 0$;
> 3. $x^2 - 5x + 4 = 0$;
> 4. $x^2 - 6x + 3 = 0$;
> 5. $x^2 - 7x + 10 = 0$.

Solution. We are going to solve the following equations by using discriminant.

1. $x^2 - 5x + 3 = 0$
   We have $\Delta = b^2 - 4ac = (-5)^2 - 4(1)(3) = 25 - 12 = 13$.
   The equation has two distinct real roots. They are
   $$x_1 = \dfrac{-b + \sqrt{\Delta}}{2a} = \dfrac{-(-5) + \sqrt{13}}{2(1)} = \dfrac{5 + \sqrt{13}}{2}$$
   and
   $$x_2 = \dfrac{-b - \sqrt{\Delta}}{2a} = \dfrac{-(-5) - \sqrt{13}}{2(1)} = \dfrac{5 - \sqrt{13}}{2}.$$

2. $2x^2 - 3x - 4 = 0$
   We have $\Delta = b^2 - 4ac = (-3)^2 - 4(2)(-4) = 9 + 32 = 41$.
   The given equation has two distinct real roots. They are
   $$x_1 = \dfrac{-b + \sqrt{\Delta}}{2a} = \dfrac{-(-3) + \sqrt{41}}{2(1)} = \dfrac{3 + \sqrt{41}}{2}$$
   and
   $$x_2 = \dfrac{-b - \sqrt{\Delta}}{2a} = \dfrac{-(-3) - \sqrt{41}}{2(1)} = \dfrac{3 - \sqrt{41}}{2}.$$

3. $x^2 - 5x + 4 = 0$
   We have $\Delta = b^2 - 4ac = (-5)^2 - 4(1)(4) = 25 - 16 = 9$.
   The equation has two distinct real roots. They are
   $$x_1 = \frac{-b + \sqrt{\Delta}}{2a} = \frac{-(-5) + \sqrt{9}}{2(1)} = \frac{5+3}{2} = \frac{8}{2} = 4$$
   and
   $$x_2 = \frac{-b - \sqrt{\Delta}}{2a} = \frac{-(-5) - \sqrt{9}}{2(1)} = \frac{5-3}{2} = \frac{2}{2} = 1.$$

4. $x^2 - 6x + 3 = 0$

   We have $\Delta = b^2 - 4ac = (-6)^2 - 4(1)(3) = 36 - 12 = 24$.
   The equation has two distinct roots. They are
   $$x_1 = \frac{-b + \sqrt{\Delta}}{2a} = \frac{-(-5) + \sqrt{24}}{2(1)} = \frac{5 + 2\sqrt{6}}{2}$$
   and
   $$x_2 = \frac{-b - \sqrt{\Delta}}{2a} = \frac{-(-5) - \sqrt{24}}{2(1)} = \frac{5 - 2\sqrt{6}}{2}.$$

5. $x^2 - 7x + 10 = 0$
   We have $\Delta = b^2 - 4ac = (-7)^2 - 4(1)(10) = 49 - 40 = 9$.
   The equation has two distinct roots. They are
   $$x_1 = \frac{-b + \sqrt{\Delta}}{2a} = \frac{-(-7) + \sqrt{9}}{2(1)} = \frac{7+3}{2} = \frac{10}{2} = 5$$
   and
   $$x_2 = \frac{-b - \sqrt{\Delta}}{2a} = \frac{-(-7) - \sqrt{9}}{2(1)} = \frac{7-3}{2} = \frac{4}{2} = 2.$$

Remark 9. There are two nice cases to consider relevant to the relation between the coefficients a, b and c.

- If $a + b + c = 0$, the equation has two roots. They are $x_1 = 1$ and $x_2 = \dfrac{c}{a}$.

9.2. How To Solve It

- If $a-b+c=0$, the equation has two roots. They are $x_1 = -1$ and $x_2 = -\dfrac{c}{a}$.

Proof. • If $a+b+c=0$, we shall prove that $x_1 = 1$ and $x_2 = \dfrac{c}{a}$.
Since $a+b+c=0$ or $b=-a-c$, then $ax^2+bx+c=0$ is equivalent to

$$ax^2 + (-a-c)x + c = 0$$
$$ax^2 - ax - cx + c = 0$$
$$ax(x-1) - c(x-1) = 0$$
$$(x-1)(ax-c) = 0.$$

We obtain $\begin{bmatrix} x-1=0 \\ ax-c=0 \end{bmatrix}$ or $\begin{bmatrix} x=1 \\ x=\dfrac{c}{a} \end{bmatrix}$.

Thus, the statement is proved.

- If $a-b+c=0$, we shall prove that $x_1 = -1$ and $x_2 = -\dfrac{c}{a}$.
Since $a-b+c=0$ or $b=a+c$, then $ax^2+bx+c=0$ is equivalent to

$$ax^2 + (a+c)x + c = 0$$
$$ax^2 + ax + cx + c = 0$$
$$ax(x+1) + c(x+1) = 0$$
$$(x+1)(ax+c) = 0.$$

We obtain $\begin{bmatrix} x+1=0 \\ ax+c=0 \end{bmatrix}$ or $\begin{bmatrix} x=-1 \\ x=-\dfrac{c}{a} \end{bmatrix}$.

Thus, the given statement is proved. $\square$

Example 56

Solve the following equations:

1. $x^2 - 3x + 2 = 0$;
2. $x^2 - 8x + 7 = 0$;
3. $x^2 - 6x + 5 = 0$;

Chapter 9. Quadratic Equations

4. $x^2 + 9x + 8 = 0$;

5. $x^2 + 13x + 12 = 0$.

Solution. Solve the following equations:

1. $x^2 - 3x + 2 = 0$
   Observe that $a + b + c = 1 - 3 + 2 = 0$, it implies that
   $$x_1 = 1 \quad \text{and} \quad x_2 = \frac{c}{a} = \frac{2}{1} = 2.$$
   Thus, $x_1 = 1$ and $x_2 = 2$ are the solutions.

2. $x^2 - 8x + 7 = 0$
   Observe that $a + b + c = 1 - 8 + 7 = 0$, it implies that
   $$x_1 = 1 \quad \text{and} \quad x_2 = \frac{c}{a} = \frac{7}{1} = 7.$$
   Thus, $x_1 = 1$ and $x_2 = 7$ are the solutions.

3. $x^2 - 6x + 5 = 0$
   Observe that $a + b + c = 1 - 6 + 5 = 0$, it implies that
   $$x_1 = 1 \quad \text{and} \quad x_2 = \frac{c}{a} = \frac{5}{1} = 5.$$
   Thus, $x_1 = 1$ and $x_2 = 5$ are the solutions.

4. $x^2 + 9x + 8 = 0$
   Observe that $a - b + c = 1 - 9 + 8 = 0$, it implies that
   $$x_1 = 1 \quad \text{and} \quad x_2 = -\frac{c}{a} = -\frac{8}{1} = -8.$$
   Thus, $x_1 = -1$ and $x_2 = -8$ are the solutions.

5. $x^2 + 13x + 12 = 0$
   Observe that $a - b + c = 1 - 13 + 12 = 0$, it implies that
   $$x_1 = -1 \quad \text{and} \quad x_2 = -\frac{c}{a} = -\frac{12}{1} = -12.$$
   Thus, $x_1 = -1$ and $x_2 = -12$ are the solutions.

Exercises

Problem 1. Solve the following equations:
1. $x^2 - 3x + 5 = 0$;
2. $x^2 - 11x + 2 = 0$;
3. $x^2 - 8x + 12 = 0$;
4. $x^2 - 4x = 4x - 2$;
5. $(x+1)(x+2) + (x+3)(x+4) = 0$.

Problem 2. Solve the following equations:
$$(x^2 + x + 3)^2 + x^2 + x + 15 = 0.$$

Problem 3. Determine the value of m such the equation
$$x^2 - (m-2)x + m + 1 = 0$$
1. has two distinct real roots
2. has two real roots with the same values
3. has no real roots.

Problem 4. Suppose that x_1 and x_2 are the two roots of the quadratic equation
$$ax^2 + bx + c = 0.$$
Prove the following equalities:
$$\begin{cases} x_1 + x_2 = -\dfrac{b}{a} \\ x_1 x_2 = \dfrac{c}{a} \end{cases}.$$

Problem 5. Let x_1 and x_2 be the roots of the equation $x^2 - 4x + 1 = 0$. Find the values of

1. $x_1^2 + x_2^2$; 2. $\dfrac{1}{x_1} + \dfrac{1}{x_2}$; 3. $x_1^3 + x_2^3$.

Problem 6. Given that α and β are the roots of the equation $x^2 - x + 8 = 0$. Evaluate the following expressions:

1. $\dfrac{\beta}{1+\alpha^2} + \dfrac{\alpha}{1+\beta^2}$; 2. $\alpha^4 + \beta^4$.

Problem 7. Let α and β be the roots of the equation

$$(x-8)(x-9) + (x-10)(x-12) = 0.$$

Find the value of $2(11-\alpha)(11-\beta)$.

Problem 8. Let α and β be the roots of the equation

$$x(x+1) + (x+1)(x+2) + (x+2)(x+3) + (x+3)(x+1) = 0.$$

Compute $(\alpha+2)(\beta+2)$.

Solutions

Problem 1. Solve the following equations:
1. $x^2 - 3x + 5 = 0$;
2. $x^2 - 11x + 2 = 0$;
3. $x^2 - 8x + 12 = 0$;
4. $x^2 - 4x = 4x - 2$;
5. $(x+1)(x+2) + (x+3)(x+4) = 0$.

Solution. Solve the following equations:
1. $x^2 - 3x + 5 = 0$
   We have
   $$\begin{aligned}\Delta &= b^2 - 4ac \\ &= (-3)^2 - 4(1)(5) \\ &= 9 - 20 \\ &= -11 < 0.\end{aligned}$$
   Therefore, the equation has no real roots.

2. $x^2 - 11x + 2 = 0$
   We have
   $$\begin{aligned}\Delta &= b^2 - 4ac \\ &= (-11)^2 - 4(1)(2) \\ &= 121 - 8 \\ &= 113.\end{aligned}$$

The equation has two distinct real roots. They are

$$x_1 = \frac{-b + \sqrt{\Delta}}{2a} = \frac{-(-11) + \sqrt{113}}{2(1)} = \frac{11 + \sqrt{113}}{2}$$

and

$$x_2 = \frac{-b - \sqrt{\Delta}}{2a} = \frac{-(-11) - \sqrt{113}}{2(1)} = \frac{11 - \sqrt{113}}{2}.$$

3. $x^2 - 8x + 12 = 0$
   We have

$$\Delta = b^2 - 4ac$$
$$= (-8)^2 - 4(1)(12)$$
$$= 64 - 48 = 16.$$

The equation has two distinct real roots. They are

$$x_1 = \frac{-b + \sqrt{\Delta}}{2a} = \frac{-(-8) + \sqrt{16}}{2(1)} = \frac{8 + 4}{2} = \frac{12}{2} = 6$$

and

$$x_2 = \frac{-b - \sqrt{\Delta}}{2a} = \frac{-(-8) - \sqrt{16}}{2(1)} = \frac{8 - 4}{2} = \frac{4}{2} = 2.$$

4. $x^2 - 4x = 4x - 2$

   We have
$$x^2 - 4x = 4x - 2$$

   or
$$x^2 - 8x + 2 = 0.$$

   It follows that
$$\Delta = b^2 - 4ac$$
$$= (-8)^2 - 4(1)(2)$$
$$= 64 - 8 = 56.$$

9.2. How To Solve It

The given equation has two distinct real roots. They are

$$x_1 = \frac{-b + \sqrt{\Delta}}{2a}$$
$$= \frac{-(-8) + \sqrt{56}}{2(1)}$$
$$= \frac{8 + 2\sqrt{14}}{2}$$
$$= \frac{2(4 + \sqrt{14})}{2}$$
$$= 4 + \sqrt{14}$$

and

$$x_2 = \frac{-b - \sqrt{\Delta}}{2a}$$
$$= \frac{-(-8) - \sqrt{56}}{2(1)}$$
$$= \frac{8 - 2\sqrt{14}}{2}$$
$$= \frac{2(4 - \sqrt{14})}{2}$$
$$= 4 - \sqrt{14}.$$

Remark 10. Given a quadratic equation $ax^2 + bx + c = 0$, where $a \neq 0$. If b is an even number, we can use Δ' to solve this equation, where $\Delta' = (b')^2 - ac$ and $b' = \dfrac{b}{2}$.

- If $\Delta' > 0$, the given equation has two distinct real roots. They are
$$x_1 = \frac{-b' + \sqrt{\Delta}}{a}$$
and
$$x_1 = \frac{-b' - \sqrt{\Delta}}{a}.$$

- If $\Delta = 0$, the given equation has two real roots with the same values. That is, $x_1 = x_2 = -\dfrac{b'}{a}$.

- If $\Delta < 0$, the given equation has no real roots.
  The given equation in Problem 1(4) can be solved by using Δ'. Here is the solution by using Δ'.
  We have $\Delta' = (b')^2 - ac = (-4)^2 - (1)(2) = 16 - 2 = 14$.
  The given equation has two distinct real roots. They are

  $$x_1 = \frac{-b' + \sqrt{\Delta'}}{a} = \frac{-(-4) + \sqrt{14}}{1} = 4 + \sqrt{14}$$

  and

  $$x_2 = \frac{-b' - \sqrt{\Delta'}}{a} = \frac{-(-4) - \sqrt{14}}{1} = 4 - \sqrt{14}.$$

5. $(x+1)(x+2) + (x+3)(x+4) = 0$
   We have
   $$(x+1)(x+2) + (x+3)(x+4) = 0$$
   or
   $$x^2 + 2x + x + 2 + x^2 + 4x + 3x + 12 = 0.$$

   Then $2x^2 + 10x + 14 = 0$ or $x^2 + 5x + 7 = 0$.
   It follows that
   $$\Delta = b^2 - 4ac$$
   $$= 5^2 - 4(1)(7)$$
   $$= 25 - 28$$
   $$= -3 < 0.$$

   Hence, the given equation has no real roots.

Problem 2. Solve the following equations:
$$\left(x^2 + x + 3\right)^2 + x^2 + x + 15 = 0.$$

Solution. Solve the equation: $\left(x^2 + x + 3\right)^2 + x^2 + x - 9 = 0$.
We have
$$\left(x^2 + x + 3\right)^2 + x^2 + x - 9 = 0$$
or
$$\left(x^2 + x + 3\right)^2 + \left(x^2 + x + 3\right) - 12 = 0.$$

9.2. How To Solve It

Let $t = x^2 + x + 3$. The given equation is equivalent to

$$t^2 + t - 12 = 0$$

or

$$(t-3)(t+4) = 0.$$

It follows that $\begin{bmatrix} t - 3 = 0 \\ t + 4 = 0 \end{bmatrix}$ or $\begin{bmatrix} t = 3 \\ t = -4 \end{bmatrix}$.

- If $t = 3$, We obtain

$$x^2 + x + 3 = 3$$

or

$$x^2 + x = 0.$$

Then $x(x+1) = 0$.
Hence, $\begin{bmatrix} x = 0 \\ x + 1 = 0 \end{bmatrix}$ or $\begin{bmatrix} x = 0 \\ x = -1 \end{bmatrix}$.

- If $t = -4$, it follows that

$$x^2 + x = -4$$

or

$$x^2 + x + 4 = 0.$$

The equation has

$$\Delta = b^2 - 4ac = 1^2 - 4(1)(4) = 1 - 16 = -15 < 0.$$

Thus, the equation has no roots.
Consequently, $x \in \{-1, 1\}$.

Problem 3. Determine the value of m such the equation

$$x^2 - (m-2)x + m + 1 = 0$$

1. has two distinct real roots
2. has two real roots with the same values
3. has no real roots.

Chapter 9. Quadratic Equations

Solution. The discriminant of the equation $x^2-(m-2)x+m+1=0$ is defined by

$$\begin{aligned}\Delta &= b^2 - 4ac \\ &= [-(m-2)]^2 - 4(1)(m+1) \\ &= m^2 - 4m + 4 - 4m - 4 \\ &= m^2 - 8m \\ &= m(m-8).\end{aligned}$$

Determine the value of m such that the equation

1. has two distinct real roots
   The given equation has two distinct real roots if and only if $\Delta > 0$.
   It follows that $m(m-8) > 0$.
   Consequently, $\begin{cases} m > 0 \\ m - 8 > 0 \end{cases}$ or $\begin{cases} m < 0 \\ m - 8 < 0 \end{cases}$.
   Then $\begin{cases} m > 0 \\ m > 8 \end{cases}$ or $\begin{cases} m < 0 \\ m < 8 \end{cases}$. That is, $m > 8$ or $m < 0$.

2. has two real roots with the same values
   The given equation has two real roots with the same values if and only if
   $$m(m-8) = 0.$$
   Then $\begin{bmatrix} m = 0 \\ m - 8 = 0 \end{bmatrix}$ or $\begin{bmatrix} m = 0 \\ m = 8 \end{bmatrix}$.
   Henec, $x \in \{0, 8\}$.

3. has no real roots
   The given equation has no real roots if and only if $\Delta < 0$.
   It follows that $m(m-8) < 0$.
   Consequently, $\begin{cases} m < 0 \\ m - 8 > 0 \end{cases}$ or $\begin{cases} m > 0 \\ m - 8 < 0 \end{cases}$.
   Then $\begin{cases} m < 0 \\ m > 8 \end{cases}$ or $\begin{cases} m > 0 \\ m < 8 \end{cases}$. That is, $0 < m < 8$.

Problem 4. Suppose that x_1 and x_2 are the two roots of the quadratic equation
$$ax^2 + bx + c = 0.$$

9.2. How To Solve It

Prove the following equalities:

$$\begin{cases} x_1 + x_2 = -\dfrac{b}{a} \\ x_1 x_2 = \dfrac{c}{a} \end{cases}.$$

Solution. Prove that $\begin{cases} x_1 + x_2 = -\dfrac{b}{a} \\ x_1 x_2 = \dfrac{c}{a} \end{cases}$.

Since x_1 and x_2 are the roots of the quadratic equation $ax^2 + bx + c = 0$, it follows that

$$\begin{aligned} ax^2 + bx + c &= a(x - x_1)(x - x_2) \\ &= a(x^2 - x_1 x - x_2 x + x_1 x_2) \\ &= ax^2 - a(x_1 + x_2)x + a x_1 x_2. \end{aligned}$$

We obtain $\begin{cases} -a(x_1 + x_2) = b \\ a x_1 x_2 = c \end{cases}$.

Consequently, $\begin{cases} x_1 + x_2 = -\dfrac{b}{a} \\ x_1 x_2 = \dfrac{c}{a} \end{cases}$.

Problem 5. Let x_1 and x_2 be the roots of the equation $x^2 - 4x + 1 = 0$. Find the values of

1. $x_1^2 + x_2^2$;
2. $\dfrac{1}{x_1} + \dfrac{1}{x_2}$;
3. $x_1^3 + x_2^3$.

Solution. Since x_1 and x_2 are the roots of the equation $x^2 - 4x + 1 = 0$, from the proof of Problem 4, we obtain $\begin{cases} x_1 + x_2 = -\dfrac{b}{a} = 4 \\ x_1 x_2 = \dfrac{c}{a} = 1 \end{cases}$.

Find the values of

1. $x_1^2 + x_2^2$

   Observe that

$$\begin{aligned} x_1^2 + x_2^2 &= x_1^2 + 2 x_1 x_2 + x_2^2 - 2 x_1 x_2 \\ &= (x_1 + x_2)^2 - 2 x_1 x_2 \\ &= 4^2 - 2(1) \\ &= 16 - 2 = 14. \end{aligned}$$

2. $\dfrac{1}{x_1} + \dfrac{1}{x_2}$
We have
$$\dfrac{1}{x_1} + \dfrac{1}{x_2} = \dfrac{x_1 + x_2}{x_1 x_2}$$
$$= \dfrac{4}{1} = 4.$$

3. $x_1^3 + x_2^3$
We have
$$\begin{aligned}x_1^3 + x_2^3 &= (x_1 + x_2)\left(x_1^2 - x_1 x_2 + x_2^2\right)\\&= (x_1 + x_2)\left[(x_1^2 + 2x_1 x_2 + x_2^2) - 3x_1 x_2\right]\\&= (x_1 + x_2)\left[(x_1 + x_2)^2 - 3x_1 x_2\right]\\&= 4\left[4^2 - 3(1)\right]\\&= 4(16 - 3)\\&= 4 \times 13 = 52.\end{aligned}$$

Problem 6. Given that α and β are the roots of the equation $x^2 - x + 8 = 0$. Evaluate the following expressions:

1. $\dfrac{\beta}{1 + \alpha^2} + \dfrac{\alpha}{1 + \beta^2}$;   2. $\alpha^4 + \beta^4$.

Solution. Since α and β are the roots of the equation $x^2 - x + 8 = 0$, it follows that
$$\alpha + \beta = -\dfrac{b}{a} = 1$$
and
$$\alpha\beta = \dfrac{c}{a} = 8.$$

Evaluate:

1. $\dfrac{\beta}{1 + \alpha^2} + \dfrac{\alpha}{1 + \beta^2}$
Observe that
$$\dfrac{\beta}{1 + \alpha^2} + \dfrac{\alpha}{1 + \beta^2} = \dfrac{\beta + \beta^3 + \alpha + \alpha^3}{(1 + \alpha^2)(1 + \beta^2)}$$

9.2. How To Solve It

$$= \frac{\alpha + \beta + (\alpha^3 + \beta^3)}{1 + \beta^2 + \alpha^2 + \alpha^2\beta^2}$$

$$= \frac{\alpha + \beta + (\alpha + \beta)(\alpha^2 - \alpha\beta + \beta^2)}{1 + (\alpha^2 + 2\alpha\beta + \beta^2) - 2\alpha\beta + \alpha^2\beta^2}$$

$$= \frac{\alpha + \beta + (\alpha + \beta)\left[(\alpha + \beta)^2 - 3\alpha\beta\right]}{1 + (\alpha + \beta)^2 - 2\alpha\beta + (\alpha\beta)^2}$$

$$= \frac{1 + (1)\left[1^2 - 3(8)\right]}{1 + 1^2 - 2(8) + 8^2}$$

$$= \frac{1 - 23}{2 - 16 + 64} = \frac{-22}{50} = -\frac{11}{25}.$$

2. $\alpha^4 + \beta^4$

    Observe that

$$\alpha^4 + \beta^4 = \left(\alpha^2\right)^2 + \left(\beta^2\right)^2 + 2\alpha^2\beta^2 - 2\alpha^2\beta^2$$
$$= \left(\alpha^2 + \beta^2\right)^2 - 2\alpha^2\beta^2$$
$$= \left[(\alpha^2 + 2\alpha\beta + \beta^2) - 2\alpha\beta\right]^2 - 2\alpha^2\beta^2$$
$$= \left[(\alpha + \beta)^2 - 2\alpha\beta\right]^2 - 2(\alpha\beta)^2$$
$$= \left[1^2 - 2(8)\right]^2 - 2(8)^2$$
$$= (1 - 16)^2 - 2(64)$$
$$= 225 - 128 = 97.$$

Problem 7. Let α and β be the roots of the equation

$$(x - 8)(x - 9) + (x - 10)(x - 12) = 0.$$

Find the value of $2(11 - \alpha)(11 - \beta)$.

Solution. Find the value of $2(11 - \alpha)(11 - \beta)$.
We have
$$(x - 8)(x - 9) + (x - 10)(x - 12) = 0$$
or
$$x^2 - 9x - 8x + 72 + x^2 - 10x - 12x + 120 = 0.$$

Then $2x^2 - 39x + 192 = 0$. It follows that $\alpha + \beta = -\dfrac{b}{a} = \dfrac{39}{2}$

and $\alpha\beta = \dfrac{c}{a} = \dfrac{192}{2} = 96$.

Observe that

$$\begin{aligned}
2(11-\alpha)(11-\beta) &= 2(121 - 11\beta - 11\alpha + \alpha\beta) \\
&= 2[121 - (\alpha+\beta)11 + \alpha\beta] \\
&= 2\left[121 - 11\left(\dfrac{39}{2}\right) + 96\right] \\
&= 2\left(217 - \dfrac{429}{2}\right) \\
&= 434 - 429 \\
&= 5.
\end{aligned}$$

Problem 8. Let α and β be the roots of the equation

$$x(x+1) + (x+1)(x+2) + (x+2)(x+3) + (x+3)(x+1) = 0.$$

Compute $(\alpha+2)(\beta+2)$.

Solution. Compute $(\alpha+2)(\beta+2)$.

We have

$$x(x+1) + (x+1)(x+2) + (x+2)(x+3) + (x+3)(x+1) = 0$$

or

$$x^2 + x + x^2 + 3x + 2 + x^2 + 5x + 6 + x^2 + 4x + 3 = 0.$$

Then $4x^2 + 13x + 11 = 0$.

It follows that $\alpha + \beta = -\dfrac{b}{a} = -\dfrac{13}{4}$ and $\alpha\beta = \dfrac{c}{a} = \dfrac{11}{4}$.

Observe that

$$\begin{aligned}
(\alpha+2)(\beta+2) &= \alpha\beta + 2\alpha + 2\beta + 4 \\
&= \alpha\beta + 2(\alpha+\beta) + 4 \\
&= \dfrac{11}{4} + 2\left(-\dfrac{13}{4}\right) + 4 \\
&= \dfrac{11}{4} - \dfrac{26}{4} + 4 \\
&= -\dfrac{15}{4} + 4 \\
&= \dfrac{1}{4}
\end{aligned}$$

Selected Problems

Problem 1. Simplify the following expressions:

1. $\sqrt{7 - 4\sqrt{3}} + \sqrt{7 + 4\sqrt{3}}$;

2. $\sqrt{2 - \sqrt{3}} + \sqrt{2 + \sqrt{3}}$;

3. $\sqrt{8 - 2\sqrt{15}} + \sqrt{8 + 2\sqrt{15}}$;

4. $\sqrt{4 + \sqrt{7}} - \sqrt{4 - \sqrt{7}}$;

5. $\sqrt{4 + \sqrt{10 + 2\sqrt{5}}} + \sqrt{4 - \sqrt{10 + 2\sqrt{5}}}$;

6. $\sqrt{1 + \sqrt{4 + \sqrt{13 + 4\sqrt{3}}}}$.

Problem 2. Prove that $\sqrt[4]{49 + 20\sqrt{6}} + \sqrt[4]{49 - 20\sqrt{6}} = 2\sqrt{3}$.

Problem 3. Prove that $\sqrt[4]{92 + 32\sqrt{7}} - \sqrt[4]{92 - 32\sqrt{7}}$ is an integer.

Problem 4. Find the value of

1. $\sqrt{5 + \sqrt{13 + \sqrt{5 + \sqrt{13 + \ldots}}}}$;

2. $\sqrt{3 + \sqrt{15 + \sqrt{3 + \sqrt{15 + \ldots}}}}$.

Chapter 9. Quadratic Equations

Problem 5. Prove that $\sqrt[3]{182+\sqrt{33125}}+\sqrt[3]{182-\sqrt{33125}}$ is an integer.

Problem 6. Let x, y and $z > 0$ such that $xy+yz+zx = 1$. Find the value of

$$S = x\sqrt{\frac{(1+y^2)(1+z^2)}{1+x^2}}+y\sqrt{\frac{(1+y^2)(1+x^2)}{1+y^2}}+x\sqrt{\frac{(1+x^2)(1+y^2)}{1+z^2}}.$$

Problem 7. Evaluate

$$S = \frac{1}{2\sqrt{1}+1\sqrt{2}}+\frac{1}{3\sqrt{2}+2\sqrt{3}}+\ldots+\frac{1}{(n+1)\sqrt{n}+n\sqrt{n+1}}.$$

Problem 8. Compute $S = \dfrac{1}{\sqrt{1}+\sqrt{2}}+\dfrac{1}{\sqrt{2}+\sqrt{3}}+\ldots+\dfrac{1}{\sqrt{n+1}+\sqrt{n}}.$

Problem 9. 1. Simplify $\sqrt{1+n^2+\dfrac{n^2}{(n+1)^2}}+\dfrac{n}{n+1}$;

2. Evaluate

$$\sqrt{\frac{1}{1^2}+\frac{1}{2^2}+\frac{1}{3^2}}+\sqrt{\frac{1}{1^2}+\frac{1}{3^2}+\frac{1}{4^2}}+\ldots+\sqrt{\frac{1}{1^2}+\frac{1}{n^2}+\frac{1}{(n+1)^2}}.$$

Problem 10. Given that a,b,c,x,y and z are nonzero distinct real numbers such that $\dfrac{x^2-yz}{a}=\dfrac{y^2-zx}{b}=\dfrac{z^2-xy}{c}$. Prove that $\dfrac{a^2-bc}{x}, \dfrac{b^2-ca}{y}$ and $\dfrac{c^2-ab}{z}$ are all positive real numbers.

Problem 11. Let x,y and z are real numbers such that $xyz = 1$. Prove that

$$\frac{1}{1+x+xy}+\frac{1}{1+y+yz}+\frac{1}{1+z+zx}=1.$$

Problem 12. Given that $x = \dfrac{a-b}{a+b}, y = \dfrac{b-c}{b+c}$ and $z = \dfrac{c-a}{c+a}$. Prove that

$$(1+x)(1+y)(1+z) = (1-x)(1-y)(1-z).$$

9.2. How To Solve It

Problem 13. Let a, b and c be three distinct real numbers. Prove that
$$\frac{a+b}{a-b}\frac{b+c}{b-c} + \frac{c+a}{c-a}\frac{b+c}{b-c} + \frac{c+a}{c-a}\frac{a+b}{a-b} = 1.$$

Problem 14. Given that $x + y = a + b$ and $x^2 + y^2 = a^2 + b^2$. Prove that $x^n + y^n = a^n + b^n$ for all natural numbers n.

Problem 15. Given that $a > b > 0$ and satisfy $a^2 + b^2 = 2025ab$. Compute $A = \dfrac{a-b}{a+b}$.

Problem 16. Given x, y and z be three real numbers that satisfy $x + y + z = 0$ and $x^2 + y^2 + z^2 = a^2$. Find $x^4 + y^4 + z^4$ in terms of a.

Problem 17. Compute
$$P = \left(1 - \frac{4}{1}\right)\left(1 - \frac{4}{9}\right)\left(1 - \frac{4}{25}\right) \cdots \left[1 - \frac{4}{(2n-1)^2}\right].$$

Problem 18. Suppose that x is a nonzero real number. Let $S_n = x^n + \dfrac{1}{x^n}$ for all nonnegative integers n. Prove that
$$S_{n+1} = S_n S_1 - S_{n-1}$$
for all $n \geq 1$.

Problem 19. Let x be a positive real number such that
$$x^2 + \frac{1}{x^2} = 2.$$

Evaluate:

1. $x + \dfrac{1}{x}$;

2. $x^3 + \dfrac{1}{x^3}$;

3. $x^5 + \dfrac{1}{x^5}$;

4. $x^7 + \dfrac{1}{x^7}$.

Problem 20. Given that a and b are two real numbers such that $ab = 1$. Prove that
$$a^{n+1} + b^{n+1} = (a^n + b^n)(a + b) - \left(a^{n-1} + b^{n-1}\right)$$
for all positive integers n.

Chapter 9. Quadratic Equations

Problem 21. Given six nonzero real numbers a, b, c, x, y and z such that $a = by + cz, b = cz + ax$ and $c = ax + by$. Prove that

$$(x+1)(y+1)(z+1) = \frac{(a+b+c)^3}{8abc}.$$

Problem 22. Given three nonzero real numbers a, b and c such that
$$\frac{1}{a} + \frac{1}{b} + \frac{1}{c} = \frac{1}{a+b+c}.$$
Prove that $a + b = 0, b + c = 0$ or $c + a = 0$.

Problem 23. Given that a, b and c are three real numbers such that $a + b + c = 0$. Prove that

$$a^4 + b^4 + c^4 = \frac{1}{2}\left(a^2 + b^2 + c^2\right)^2.$$

Solutions

Problem 1. Simplify the following expressions:

1. $\sqrt{7 - 4\sqrt{3}} + \sqrt{7 + 4\sqrt{3}}$;

2. $\sqrt{2 - \sqrt{3}} + \sqrt{2 + \sqrt{3}}$;

3. $\sqrt{8 - 2\sqrt{15}} + \sqrt{8 + 2\sqrt{15}}$;

4. $\sqrt{4 + \sqrt{7}} - \sqrt{4 - \sqrt{7}}$;

5. $\sqrt{4 + \sqrt{10 + 2\sqrt{5}}} + \sqrt{4 - \sqrt{10 + 2\sqrt{5}}}$;

6. $\sqrt{1 + \sqrt{4 + \sqrt{13 + 4\sqrt{3}}}}$.

Solution. Simplify the following expressions:

1. $\sqrt{7 - 4\sqrt{3}} + \sqrt{7 + 4\sqrt{3}}$
   We have

$$\sqrt{7 - 4\sqrt{3}} + \sqrt{7 + 4\sqrt{3}}$$
$$= \sqrt{4 - 4\sqrt{3} + 3} + \sqrt{4 + 4\sqrt{3} + 3}$$
$$= \sqrt{2^2 - 2(2)\left(2\sqrt{3}\right) + \sqrt{3^2}} + \sqrt{2^2 + 2(2)\left(2\sqrt{3}\right) + \sqrt{3^2}}$$
$$= \sqrt{\left(2 - \sqrt{3}\right)^2} + \sqrt{\left(2 + \sqrt{3}\right)^2}$$

Chapter 9. Quadratic Equations

$$= 2 - \sqrt{3} + 2 + \sqrt{3}$$
$$= 4.$$

Therefore, $\sqrt{7 - 4\sqrt{3}} + \sqrt{7 + 4\sqrt{3}} = 4$.

2. $\sqrt{2 - \sqrt{3}} + \sqrt{2 + \sqrt{3}}$
We have

$$\sqrt{2 - \sqrt{3}} + \sqrt{2 + \sqrt{3}}$$
$$= \sqrt{\frac{4 - 2\sqrt{3}}{2}} + \sqrt{\frac{4 + 2\sqrt{3}}{2}}$$
$$= \frac{\sqrt{4 - 2\sqrt{3}}}{\sqrt{2}} + \frac{\sqrt{4 + 2\sqrt{3}}}{\sqrt{2}}$$
$$= \frac{\sqrt{3 - 2\sqrt{3} + 1}}{\sqrt{2}} + \frac{\sqrt{3 + 2\sqrt{3} + 1}}{\sqrt{2}}$$
$$= \frac{\sqrt{\sqrt{3}^2 - 2\left(\sqrt{3}\right)(1) + 1^2}}{\sqrt{2}} + \frac{\sqrt{\sqrt{3}^2 + 2\left(\sqrt{3}\right)(1) + 1^2}}{\sqrt{2}}$$
$$= \frac{\sqrt{\left(\sqrt{3} - 1\right)^2}}{\sqrt{2}} + \frac{\sqrt{\left(\sqrt{3} + 1\right)^2}}{\sqrt{2}}$$
$$= \frac{\sqrt{3} - 1}{\sqrt{2}} + \frac{\sqrt{3} + 1}{\sqrt{2}}$$
$$= \frac{2\sqrt{3}}{\sqrt{2}}$$
$$= \frac{2\sqrt{6}}{2}$$
$$= \sqrt{6}.$$

Therefore, $\sqrt{2 - \sqrt{3}} + \sqrt{2 + \sqrt{3}} = \sqrt{6}$.

3. $\sqrt{8 - 2\sqrt{15}} + \sqrt{8 + 2\sqrt{15}}$
We have

$$\sqrt{8 - 2\sqrt{15}} + \sqrt{8 + 2\sqrt{15}}$$

9.2. How To Solve It

$$= \sqrt{5 - 2\sqrt{15} + 3} + \sqrt{5 + 2\sqrt{15} + 3}$$
$$= \sqrt{\sqrt{5}^2 - 2\left(\sqrt{5}\right)\left(\sqrt{3}\right) + \sqrt{3}^2}$$
$$+ \sqrt{\sqrt{5}^2 + 2\left(\sqrt{5}\right)\left(\sqrt{3}\right) + \sqrt{3}^2}$$
$$= \sqrt{\left(\sqrt{5} - \sqrt{3}\right)^2} + \sqrt{\left(\sqrt{5} + \sqrt{3}\right)^2}$$
$$= \sqrt{5} - \sqrt{3} + \sqrt{5} + \sqrt{3}$$
$$= 2\sqrt{5}.$$

Therefore, $\sqrt{8 - 2\sqrt{15}} + \sqrt{8 + 2\sqrt{15}} = 2\sqrt{5}.$

4. $\sqrt{4 + \sqrt{7}} - \sqrt{4 - \sqrt{7}}$
   We have

$$\sqrt{4 + \sqrt{7}} - \sqrt{4 - \sqrt{7}}$$
$$= \sqrt{\frac{8 + 2\sqrt{7}}{2}} - \sqrt{\frac{8 - 2\sqrt{7}}{2}}$$
$$= \frac{\sqrt{8 + 2\sqrt{7}}}{\sqrt{2}} - \frac{\sqrt{8 - 2\sqrt{7}}}{\sqrt{2}}$$
$$= \frac{\sqrt{7 + 2\sqrt{7} + 1}}{\sqrt{2}} - \frac{\sqrt{7 - 2\sqrt{7} + 1}}{\sqrt{2}}$$
$$= \frac{\sqrt{\sqrt{7}^2 + 2\sqrt{7} + 1^2}}{\sqrt{2}} - \frac{\sqrt{\sqrt{7}^2 - 2\sqrt{7} + 1^2}}{\sqrt{2}}$$
$$= \frac{\sqrt{\left(\sqrt{7} + 1\right)^2}}{\sqrt{2}} - \frac{\sqrt{\left(\sqrt{7} - 1\right)^2}}{\sqrt{2}}$$
$$= \frac{\sqrt{7} + 1 - \left(\sqrt{7} - 1\right)}{\sqrt{2}}$$
$$= \frac{\sqrt{7} + 1 - \sqrt{7} + 1}{\sqrt{2}}$$
$$= \frac{2}{\sqrt{2}}$$

$$= \frac{2\sqrt{2}}{2}$$
$$= \sqrt{2}.$$

5. $\sqrt{4+\sqrt{10+2\sqrt{5}}} + \sqrt{4-\sqrt{10+2\sqrt{5}}}$
We have

$$A^2 = \left(\sqrt{4+\sqrt{10+2\sqrt{5}}} + \sqrt{4-\sqrt{10+2\sqrt{5}}}\right)^2$$
$$= 4 + \sqrt{10+2\sqrt{5}} + 2\sqrt{16 - \sqrt{10+2\sqrt{5}}^2}$$
$$\quad + 4 - \sqrt{10+2\sqrt{5}}$$
$$= 8 + 2\sqrt{16 - 10 - 2\sqrt{5}}$$
$$= 8 + 2\sqrt{6 - 2\sqrt{5}}$$
$$= 8 + 2\sqrt{5 - 2\sqrt{5} + 1}$$
$$= 8 + 2\sqrt{\sqrt{5}^2 - 2\sqrt{5} + 1^2}$$
$$= 8 + 2\sqrt{\left(\sqrt{5}-1\right)^2}$$
$$= 8 + 2\left(\sqrt{5}-1\right)$$
$$= 8 + 2\sqrt{5} - 2$$
$$= 6 + 2\sqrt{5}$$
$$= 5 + 2\sqrt{5} + 1$$
$$= \sqrt{5}^2 + 2\sqrt{5} + 1^2$$
$$= \left(\sqrt{5}+1\right)^2.$$

Since $A > 0$, it follows that $A = \sqrt{5} + 1$.

6. $\sqrt{1+\sqrt{4+\sqrt{13+4\sqrt{3}}}}$

9.2. How To Solve It

We have

$$\sqrt{1+\sqrt{3+\sqrt{13+4\sqrt{3}}}}$$

$$=\sqrt{1+\sqrt{3+\sqrt{13+2\left(2\sqrt{3}\right)}}}$$

$$=\sqrt{1+\sqrt{3+\sqrt{\left(2\sqrt{3}\right)^2+2\left(2\sqrt{3}\right)(1)+1^2}}}$$

$$=\sqrt{1+\sqrt{3+\sqrt{\left(2\sqrt{3}+1\right)^2}}}$$

$$=\sqrt{1+\sqrt{3+2\sqrt{3}+1}}$$

$$=\sqrt{1+\sqrt{4+2\sqrt{3}}}$$

$$=\sqrt{1+\sqrt{3+2\sqrt{3}+1}}$$

$$=\sqrt{1+\sqrt{\sqrt{3^2}+2\sqrt{3}+1}}$$

$$=\sqrt{1+\sqrt{\left(\sqrt{3}+1\right)^2}}$$

$$=\sqrt{1+\sqrt{3}+1}$$

$$=\sqrt{2+\sqrt{3}}$$

$$=\sqrt{\frac{4+2\sqrt{3}}{2}}$$

$$=\frac{\sqrt{3+2\sqrt{3}+1}}{\sqrt{2}}$$

$$=\frac{\sqrt{\left(\sqrt{3}+1\right)^2}}{\sqrt{2}}$$

$$=\frac{\sqrt{3}+1}{\sqrt{2}}$$

$$= \frac{\sqrt{6}+\sqrt{2}}{2}.$$

Therefore, $\sqrt{1+\sqrt{4+\sqrt{13+4\sqrt{3}}}} = \dfrac{\sqrt{6}+\sqrt{2}}{2}.$

Problem 2. Prove that $\sqrt[4]{49+20\sqrt{6}} + \sqrt[4]{49-20\sqrt{6}} = 2\sqrt{3}.$

Solution. Prove that $\sqrt[4]{49+20\sqrt{6}} + \sqrt[4]{49-20\sqrt{6}} = 2\sqrt{3}.$
We have

$$\sqrt[4]{49+20\sqrt{6}} + \sqrt[4]{49-20\sqrt{6}}$$
$$= \sqrt[4]{25+20\sqrt{6}+24} + \sqrt[4]{25-20\sqrt{6}+24}$$
$$= \sqrt[4]{5^2 + 2(5)\left(2\sqrt{6}\right) + \left(2\sqrt{6}\right)^2} + \sqrt[4]{5^2 - 2(5)\left(2\sqrt{6}\right) + \left(2\sqrt{6}\right)^2}$$
$$= \sqrt[4]{\left(5+2\sqrt{6}\right)^2} + \sqrt[4]{\left(5-2\sqrt{6}\right)^2}$$
$$= \sqrt{5+2\sqrt{6}} + \sqrt{5-2\sqrt{6}}$$
$$= \sqrt{\sqrt{3}^2 + 2\sqrt{6} + \sqrt{2}^2} + \sqrt{\sqrt{3}^2 - 2\sqrt{6} + \sqrt{2}^2}$$
$$= \sqrt{\left(\sqrt{3}+\sqrt{2}\right)^2} + \sqrt{\left(\sqrt{3}-\sqrt{2}\right)^2}$$
$$= \sqrt{3} + \sqrt{2} + \sqrt{3} - \sqrt{2}$$
$$= 2\sqrt{3}.$$

Therefore, $\sqrt[4]{49+20\sqrt{6}} + \sqrt[4]{49-20\sqrt{6}} = 2\sqrt{3}.$

Problem 3. Prove that $\sqrt[4]{92+32\sqrt{7}} - \sqrt[4]{92-32\sqrt{7}}$ is an integer.

Solution. Prove that $\sqrt[4]{92+32\sqrt{7}} - \sqrt[4]{92-32\sqrt{7}}$ is an integer.
We have

$$\sqrt[4]{92+32\sqrt{7}} - \sqrt[4]{92-32\sqrt{7}}$$
$$= \sqrt[4]{8^2 + 2(8)\left(2\sqrt{7}\right) + \left(2\sqrt{7}\right)^2} - \sqrt[4]{8^2 - 2(8)\left(2\sqrt{7}\right) + \left(2\sqrt{7}\right)^2}$$

9.2. How To Solve It

$$= \sqrt[4]{\left(8+2\sqrt{7}\right)^2} - \sqrt[4]{\left(8-2\sqrt{7}\right)^2}$$
$$= \sqrt{8+2\sqrt{7}} - \sqrt{8-2\sqrt{7}}$$
$$= \sqrt{7+2\sqrt{7}+1} - \sqrt{7-2\sqrt{7}+1}$$
$$= \sqrt{\sqrt{7}^2 + 2\sqrt{7} + 1^2} - \sqrt{\sqrt{7}^2 - 2\sqrt{7} + 1^2}$$
$$= \sqrt{\left(\sqrt{7}+1\right)^2} - \sqrt{\left(\sqrt{7}-1\right)^2}$$
$$= \sqrt{7}+1 - \left(\sqrt{7}-1\right)$$
$$= \sqrt{7}+1 - \sqrt{7}+1$$
$$= 2.$$

Therefore, $\sqrt[4]{92+32\sqrt{7}} - \sqrt[4]{92-32\sqrt{7}}$ is an integer.

Problem 4. Find the value of

1. $\sqrt{5+\sqrt{13+\sqrt{5+\sqrt{13+...}}}}$;

2. $\sqrt{3+\sqrt{15+\sqrt{3+\sqrt{15+...}}}}$.

Solution. Find the value of

1. $\sqrt{5+\sqrt{13+\sqrt{5+\sqrt{13+...}}}}$

Let $x = \sqrt{5+\sqrt{13+\sqrt{5+\sqrt{13+...}}}}$.

Then $x^2 = 5 + \sqrt{13+\sqrt{5+\sqrt{13+...}}}$.

It follows that

$$x^2 - 5 = \sqrt{13+\sqrt{5+\sqrt{13+...}}}$$

or

$$\left(x^2 - 5\right)^2 = 13 + \sqrt{5+\sqrt{13+\sqrt{5+\sqrt{13+...}}}} = 13 + x.$$

Consequently,
$$x^4 - 10x^2 + 25 = 13 + x$$
$$x^4 - 10x^2 - x + 12 = 0$$
$$(x^4 - 9x^2) - (x^2 - 9) - (x - 3) = 0$$
$$x^2(x^2 - 9) - (x - 3)(x + 3) - (x - 3) = 0$$
$$x^2(x - 3)(x + 3) - (x - 3)(x + 3) - (x - 3) = 0$$
$$(x - 3)\left[x^2(x + 3) - (x + 3) - 1\right] = 0$$
$$(x - 3)\left[(x + 3)(x^2 - 1) - 1\right] = 0$$
$$(x - 3)\left[(x + 3)(x - 1)(x + 1) - 1\right] = 0.$$

Moreover, $x = \sqrt{5 + \sqrt{13 + \sqrt{5 + \sqrt{13 + \ldots}}}} > 2$.
Then $(x + 3)(x - 1)(x + 1) - 1 > 0$.
Hence, $x - 3 = 0$ or $x = 3$.
Thus, $\sqrt{5 + \sqrt{13 + \sqrt{5 + \sqrt{13 + \ldots}}}} = 3$.

2. $\sqrt{3 + \sqrt{33 + \sqrt{3 + \sqrt{33 + \ldots}}}}$

Let $x = \sqrt{3 + \sqrt{33 + \sqrt{3 + \sqrt{33 + \ldots}}}}$.
Squaring both sides of the equation, we obtain
$$x^2 = 3 + \sqrt{33 + \sqrt{3 + \sqrt{33 + \sqrt{3 + \ldots}}}}.$$

It follows that
$$x^2 - 3 = \sqrt{33 + \sqrt{3 + \sqrt{33 + \sqrt{3 + \ldots}}}}.$$

It implies that
$$(x^2 - 3)^2 = 33 + x.$$
Then
$$x^4 - 6x^2 + 9 = 33 + x.$$

9.2. How To Solve It

We obtain

$$x^4 - 6x^2 + 9 = 33 + x$$
$$x^4 - 6x^2 - x - 24 = 0$$
$$x^4 - 3x^3 + 3x^3 - 9x^2 + 3x^2 - 9x + 8x - 24 = 0$$
$$x^3(x-3) + 3x^2(x-3) + 3x(x-3) + 8(x-3) = 0$$
$$(x-3)(x^3 + 3x^2 + 3x + 8) = 0.$$

Since $x > 0$, it follows that $x^3 + 3x^2 + 3x + 8 > 0$. It turns out that $x - 3 = 0$. Hence, $x = 3$.

Therefore, $\sqrt{3 + \sqrt{33 + \sqrt{3 + \sqrt{33 + \ldots}}}} = 3$.

Problem 5. Prove that $\sqrt[3]{182 + \sqrt{33125}} + \sqrt[3]{182 - \sqrt{33125}}$ is an integer.

Solution. Let $A = \sqrt[3]{182 + \sqrt{33125}} + \sqrt[3]{182 - \sqrt{33125}} > 0$. Using $(a+b)^3 = a^3 + b^3 + 3ab(a+b)$, it implies that

$$A^3 = 182 + \sqrt{33125} + 182 - \sqrt{33125}$$
$$+ 3\sqrt[3]{\left(182 + \sqrt{33125}\right)\left(182 - \sqrt{33125}\right)}$$
$$\times \left(\sqrt[3]{182 + \sqrt{33125}} + \sqrt[3]{182 - \sqrt{33125}}\right)$$
$$= 364 + 3\sqrt[3]{33124 - 33125}A$$
$$= 364 - 3A.$$

Then

$$A^3 + 3A - 364 = 0$$
$$A^3 - 7A^2 + 7A^2 - 49A + 52A - 364 = 0$$
$$A^2(A-7) + 7A(A-7) + 52(A-7) = 0$$
$$(A-7)(A^2 + 7A + 52) = 0.$$

Since $A^2 + 7A + 52 > 0$, it follows that $A - 7 = 0$ or $A = 7$. Hence, $\sqrt[3]{182 + \sqrt{33125}} + \sqrt[3]{182 - \sqrt{33125}} = 7$ is an integer.

Chapter 9. Quadratic Equations

Problem 6. Let x, y and $z > 0$ such that $xy + yz + zx = 1$. Find the value of

$$S = x\sqrt{\frac{(1+y^2)(1+z^2)}{1+x^2}} + y\sqrt{\frac{(1+y^2)(1+x^2)}{1+y^2}} + x\sqrt{\frac{(1+x^2)(1+y^2)}{1+z^2}}.$$

Solution. Find the value of S.
Since $xy + yz + zx = 1$, it implies that

$$1 + x^2 = xy + yz + zx + x^2$$
$$= y(x+z) + x(x+z)$$
$$= (x+z)(x+y).$$

Similarly,
$$1 + y^2 = (y+x)(y+z)$$

and
$$1 + z^2 = (z+x)(z+y).$$

We obtain
$$\frac{(1+y^2)(1+z^2)}{1+x^2} = \frac{(y+x)(y+z)(z+x)(z+y)}{(x+y)(x+z)}$$
$$= (y+z)^2.$$

Then $\sqrt{\dfrac{(1+y^2)(1+z^2)}{1+x^2}} = \sqrt{(y+z)^2} = y+z.$

Likewise,
$$\sqrt{\frac{(1+z^2)(1+x^2)}{1+y^2}} = z+x$$

and
$$\sqrt{\frac{(1+x^2)(1+y^2)}{1+z^2}} = x+y.$$

We obtain

$$S = x\sqrt{\frac{(1+y^2)(1+z^2)}{1+x^2}} + y\sqrt{\frac{(1+z^2)(1+x^2)}{1+y^2}} + z\sqrt{\frac{(1+x^2)(1+y^2)}{1+z^2}}$$
$$= x(y+z) + y(z+x) + z(x+y)$$
$$= xy + xz + yz + xy + zx + yz$$
$$= 2(xy + yz + zx)$$

9.2. How To Solve It

$= 2$.

Therefore, $S = 2$.

Problem 7. Evaluate
$$S = \frac{1}{2\sqrt{1}+1\sqrt{2}} + \frac{1}{3\sqrt{2}+2\sqrt{3}} + \ldots + \frac{1}{(n+1)\sqrt{n}+n\sqrt{n+1}}.$$

Solution. Simplify S.
For all positive integers k, we have

$$\frac{1}{(k+1)\sqrt{k}+k\sqrt{k+1}}$$
$$= \frac{(k+1)\sqrt{k}-k\sqrt{k+1}}{\left[(k+1)\sqrt{k}+k\sqrt{k+1}\right]\left[(k+1)\sqrt{k}-k\sqrt{k+1}\right]}$$
$$= \frac{(k+1)\sqrt{k}-k\sqrt{k+1}}{(k+1)^2\sqrt{k^2}-k^2\sqrt{(k+1)^2}}$$
$$= \frac{(k+1)\sqrt{k}-k\sqrt{k+1}}{k(k+1)^2-k^2(k+1)}$$
$$= \frac{(k+1)\sqrt{k}-k\sqrt{k+1}}{k(k+1)(k+1-k)}$$
$$= \frac{(k+1)\sqrt{k}-k\sqrt{k+1}}{k(k+1)}$$
$$= \frac{(k+1)\sqrt{k}}{k(k+1)} - \frac{k\sqrt{k+1}}{k(k+1)}$$
$$= \frac{\sqrt{k}}{k} - \frac{\sqrt{k+1}}{k+1}.$$

It follows that
$$\frac{1}{2\sqrt{1}+1\sqrt{2}} = \frac{1}{\sqrt{1}} - \frac{1}{\sqrt{2}};$$
$$\frac{1}{3\sqrt{2}+2\sqrt{3}} = \frac{1}{\sqrt{2}} - \frac{1}{\sqrt{3}};$$
$$\frac{1}{4\sqrt{3}+3\sqrt{4}} = \frac{1}{\sqrt{3}} - \frac{1}{\sqrt{4}};$$
$$\vdots$$

Chapter 9. Quadratic Equations

and $\dfrac{1}{(n+1)\sqrt{n}+n\sqrt{n+1}} = \dfrac{1}{\sqrt{n}} - \dfrac{1}{\sqrt{n+1}}.$

Adding all of the above equalities, we obtain

$$S = \dfrac{1}{\sqrt{1}} - \dfrac{1}{\sqrt{n+1}} = 1 - \dfrac{1}{\sqrt{n+1}}.$$

Problem 8. Compute $S = \dfrac{1}{\sqrt{1}+\sqrt{2}} + \dfrac{1}{\sqrt{2}+\sqrt{3}} + ... + \dfrac{1}{\sqrt{n+1}+\sqrt{n}}.$

Solution. Compute S.
For all positive integers k, we have

$$\dfrac{1}{\sqrt{k}+\sqrt{k+1}} = \dfrac{1}{\sqrt{k+1}+\sqrt{k}}$$
$$= \dfrac{\sqrt{k+1}-\sqrt{k}}{\left(\sqrt{k+1}+\sqrt{k}\right)\left(\sqrt{k+1}-\sqrt{k}\right)}$$
$$= \dfrac{\sqrt{k+1}-\sqrt{k}}{\sqrt{(k+1)^2}-\sqrt{k^2}}$$
$$= \dfrac{\sqrt{k+1}-\sqrt{k}}{k+1-k}$$
$$= \sqrt{k+1}-\sqrt{k}.$$

It follows that

$$\dfrac{1}{\sqrt{1}+\sqrt{2}} = \sqrt{2}-\sqrt{1};$$
$$\dfrac{1}{\sqrt{3}+\sqrt{2}} = \sqrt{3}-\sqrt{2};$$
$$\dfrac{1}{\sqrt{4}+\sqrt{3}} = \sqrt{4}-\sqrt{3};$$
$$\vdots$$

and $\dfrac{1}{\sqrt{n}+\sqrt{n+1}} = \sqrt{n+1}-\sqrt{n}.$

Adding all of the above equalities, we obtan

$$S = \sqrt{n+1}-\sqrt{1}$$
$$= \sqrt{n+1}-1.$$

9.2. How To Solve It

Problem 9. 1. Simplify $\sqrt{1 + n^2 + \dfrac{n^2}{(n+1)^2}} + \dfrac{n}{n+1}$;

2. Evaluate

$$\sqrt{\dfrac{1}{1^2} + \dfrac{1}{2^2} + \dfrac{1}{3^2}} + \sqrt{\dfrac{1}{1^2} + \dfrac{1}{3^2} + \dfrac{1}{4^2}} + \ldots + \sqrt{\dfrac{1}{1^2} + \dfrac{1}{n^2} + \dfrac{1}{(n+1)^2}}.$$

Solution. 1. Simplify $\sqrt{1 + n^2 + \dfrac{n^2}{(n+1)^2}} - \dfrac{n}{n+1}$

We have

$$\sqrt{1 + \dfrac{1}{n^2} + \dfrac{1}{(n+1)^2}} - \dfrac{n}{n+1}$$

$$= \sqrt{1 + \dfrac{1}{n^2} + \dfrac{1}{(n+1)^2} - \dfrac{2}{n+1} + \dfrac{2}{n} - \dfrac{2}{n} + \dfrac{2}{n+1}} - \dfrac{n}{n+1}$$

$$= \sqrt{1 + \dfrac{1}{n^2} + \dfrac{1}{(n+1)^2} - \dfrac{2}{n+1} + \dfrac{2}{n} - \dfrac{2}{n(n+1)}} - \dfrac{n}{n+1}$$

$$= \sqrt{\left(1 + \dfrac{1}{n} - \dfrac{1}{n+1}\right)^2} - \dfrac{n}{n+1}$$

$$= 1 + \dfrac{1}{n} - \dfrac{1}{n+1} - \dfrac{n}{n+1}$$

$$= 1 + \dfrac{1}{n} - \dfrac{n+1}{n+1}$$

$$= 1 + \dfrac{1}{n} - 1$$

$$= \dfrac{1}{n}.$$

Consequently, $\sqrt{1 + n^2 + \dfrac{n^2}{(n+1)^2}} - \dfrac{n}{n+1} = \dfrac{1}{n}$

2. From (1), $\sqrt{1 + \dfrac{1}{k^2} + \dfrac{1}{(k+1)^2}} = 1 + \dfrac{1}{k} - \dfrac{1}{k+1}$. It follows that

$$\sqrt{\dfrac{1}{1^2} + \dfrac{1}{2^2} + \dfrac{1}{3^2}} = 1 + \dfrac{1}{2} - \dfrac{1}{3};$$

$$\sqrt{\frac{1}{1^2}+\frac{1}{3^2}+\frac{1}{4^2}}=1+\frac{1}{3}-\frac{1}{4};$$

$$\sqrt{\frac{1}{1^2}+\frac{1}{4^2}+\frac{1}{5^2}}=1+\frac{1}{4}-\frac{1}{5};$$

$$\vdots$$

and $\sqrt{\dfrac{1}{1^2}+\dfrac{1}{n^2}+\dfrac{1}{(n+1)^2}}=1+\dfrac{1}{n}-\dfrac{1}{n+1}.$

Adding all of the above equalities, we obtain

$$S = n + \frac{1}{2} - \frac{1}{n+1}$$
$$= \frac{2n(n+1)+n+1-2}{2(n+1)}$$
$$= \frac{2n^2+2n+n-1}{2(n+1)}$$
$$= \frac{2n^2+3n-1}{2(n+1)}.$$

Problem 10. Given that a, b, c, x, y and z are nonzero distinct real numbers such that $\dfrac{x^2-yz}{a}=\dfrac{y^2-zx}{b}=\dfrac{z^2-xy}{c}$. Prove that $\dfrac{a^2-bc}{x}, \dfrac{b^2-ca}{y}$ and $\dfrac{c^2-ab}{z}$ are all positive real numbers.

Solution. Prove that $\dfrac{a^2-bc}{x}, \dfrac{b^2-ca}{y}$ and $\dfrac{c^2-ab}{z}$ are all positive real numbers.
Let $\dfrac{x^2-yz}{a}=\dfrac{y^2-zx}{b}=\dfrac{z^2-xy}{c}=k.$
It follows that $a=\dfrac{x^2-yz}{k}, b=\dfrac{y^2-zx}{k}$ and $c=\dfrac{z^2-xy}{k}.$
We have

$$a^2 - bc = \left(\frac{x^2-yz}{k}\right)^2 - \left(\frac{y^2-zx}{k}\right)\left(\frac{z^2-xy}{k}\right)$$
$$= \frac{x^4-2x^2yz+y^2z^2}{k^2} - \frac{y^2z^2-xy^3-xz^3+x^2yz}{k^2}$$
$$= \frac{x^4-2x^2yz+y^2z^2-y^2z^2+xy^3+xz^3-x^2yz}{k^2}$$

9.2. How To Solve It

$$= \frac{x^4 - 3x^2yz + xy^3 + xz^3}{k^2}$$
$$= \frac{x\left(x^3 + y^3 + z^3 - 3xyz\right)}{k^2}.$$

Then
$$\frac{a^2 - bc}{x} = \frac{x^3 + y^3 + z^3 - 3xyz}{k^2}.$$

Since a, b, c, x, y and z are nonzero distinct real numbers it implies that $k^2 > 0$ and $x^3 + y^3 + z^3 > 3xyz$. It turns out that $\dfrac{a^2 - bc}{x} > 0$. Similarly, we obtain
$$\frac{b^2 - ca}{y} > 0$$

and
$$\frac{c^2 - ab}{z} > 0.$$

Therefore, $\dfrac{a^2 - bc}{x}, \dfrac{b^2 - ca}{y}$ and $\dfrac{c^2 - ab}{z}$ are all positive real numbers.

Problem 11. Let x, y and z are real numbers such that $xyz = 1$. Prove that
$$\frac{1}{1+x+xy} + \frac{1}{1+y+yz} + \frac{1}{1+z+zx} = 1.$$

Solution. Prove that $\dfrac{1}{1+x+xy} + \dfrac{1}{1+y+yz} + \dfrac{1}{1+z+zx} = 1.$
We have
$$\frac{1}{1+x+xy} + \frac{1}{1+y+yz} + \frac{1}{1+z+zx}$$
$$= \frac{z}{z+zx+xyz} + \frac{xz}{xz+xyz+xyz^2} + \frac{1}{1+z+zx}$$
$$= \frac{z}{z+zx+1} + \frac{xz}{xz+1+z} + \frac{1}{1+z+zx}$$
$$= \frac{1+z+zx}{1+z+zx} = 1.$$

Therefore, $\dfrac{1}{1+x+xy} + \dfrac{1}{1+y+yz} + \dfrac{1}{1+z+zx} = 1.$

Problem 12. Given that $x = \dfrac{a-b}{a+b}, y = \dfrac{b-c}{b+c}$ and $z = \dfrac{c-a}{c+a}$.
Prove that

$$(1+x)(1+y)(1+z) = (1-x)(1-y)(1-z).$$

Solution. Prove that $(1+x)(1+y)(1+z) = (1-x)(1-y)(1-z)$.
We have

$$1 + x = 1 + \frac{a-b}{a+b} = \frac{a+b+a-b}{a+b} = \frac{2a}{a+b},$$

$$1 + y = 1 + \frac{b-c}{b+c} = \frac{b+c+b-c}{b+c} = \frac{2b}{b+c},$$

and $\quad 1 + z = 1 + \dfrac{c-a}{c+a} = \dfrac{c+a+c-a}{c+a} = \dfrac{2c}{c+a}.$

It follows that $(1+x)(1+y)(1+z) = \left(\dfrac{2a}{a+b}\right)\left(\dfrac{2b}{b+c}\right)\left(\dfrac{2c}{c+a}\right)$
or

$$(1+x)(1+y)(1+z) = \frac{8abc}{(a+b)(b+c)(c+a)}. \qquad (1)$$

Moreover,

$$1 - x = 1 - \frac{a-b}{a+b} = \frac{a+b-a+b}{a+b} = \frac{2b}{a+b};$$

$$1 - y = 1 - \frac{b-c}{b+c} = \frac{b+c-b+c}{b+c} = \frac{2c}{b+c};$$

and $\quad 1 - z = 1 - \dfrac{c-a}{c+a} = \dfrac{c+a-c+a}{c+a} = \dfrac{2a}{c+a}.$

We obtain $(1-x)(1-y)(1-z) = \left(\dfrac{2b}{a+b}\right)\left(\dfrac{2c}{b+c}\right)\left(\dfrac{2a}{c+a}\right)$
or

$$(1-x)(1-y)(1-z) = \frac{8abc}{(a+b)(b+c)(c+a)}. \qquad (2)$$

From (1) and (2), we obtain

$$(1+x)(1+y)(1+z) = (1-x)(1-y)(1-z).$$

Problem 13. Let a, b and c be three distinct real numbers. Prove that

$$\frac{a+b}{a-b}\frac{b+c}{b-c} + \frac{c+a}{c-a}\frac{b+c}{b-c} + \frac{c+a}{c-a}\frac{a+b}{a-b} = 1.$$

9.2. How To Solve It

Solution. Prove that $\dfrac{a+b}{a-b}\dfrac{b+c}{b-c} + \dfrac{c+a}{c-a}\dfrac{b+c}{b-c} + \dfrac{c+a}{c-a}\dfrac{a+b}{a-b} = 1$.

Let $x = \dfrac{a+b}{a-b}, y = \dfrac{b+c}{b-c}$ and $z = \dfrac{c+a}{c-a}$.
We obtain

$$1 + x = 1 + \frac{a+b}{a-b} = \frac{a-b+a+b}{a-b} = \frac{2a}{a-b};$$

$$1 + y = 1 + \frac{b+c}{b-c} = \frac{b-c+b+c}{b-c} = \frac{2b}{b-c};$$

and $\quad 1 + z = 1 + \dfrac{c+a}{c-a} = \dfrac{c-a+c+a}{c-a} = \dfrac{2c}{c-a}.$

It follows that $(1+x)(1+y)(1+z) = \left(\dfrac{2a}{a-b}\right)\left(\dfrac{2b}{b-c}\right)\left(\dfrac{2c}{c-a}\right)$

$$(1+x)(1+y)(1+z) = \frac{8abc}{(a-b)(b-c)(c-a)}. \qquad (1)$$

Moreover,

$$x - 1 = \frac{a+b}{a-b} - 1 = \frac{a+b-a+b}{a-b} = \frac{2b}{a-b};$$

$$y - 1 = \frac{b+c}{b-c} - 1 = \frac{b+c-b+c}{b-c} = \frac{2c}{b-c};$$

and $\quad z - 1 = \dfrac{c+a}{c-a} - 1 = \dfrac{c+a-c+a}{c-a} = \dfrac{2a}{c-a}.$

We obtain $(x-1)(y-1)(z-1) = \left(\dfrac{2a}{a-b}\right)\left(\dfrac{2b}{b-c}\right)\left(\dfrac{2c}{c-a}\right)$

$$(x-1)(y-1)(z-1) = \frac{8abc}{(a-b)(b-c)(c-a)}. \qquad (2)$$

From (1) and (2), it implies that

$$(x+1)(y+1)(z+1) = (x-1)(y-1)(z-1)$$
$$(1+x)(1+y)(1+z) = -(1-x)(1-y)(1-z)$$
$$1 + (x+y+z) + (xy+yz+zx) + xyz$$
$$= -[1 - (x+y+z) + (xy+yz+zx) - xyz]$$
$$1 + (x+y+z) + (xy+yz+zx) + xyz$$
$$= -1 + (x+y+z) - (xy+yz+zx) + xyz$$

237

$$2(xy + yz + zx) = -2$$
$$xy + yz + zx = -1.$$

Therefore, $\dfrac{a+b\,b+c}{a-b\,b-c} + \dfrac{c+a\,b+c}{c-a\,b-c} + \dfrac{c+a\,a+b}{c-a\,a-b} = 1.$

Problem 14. Given that $x + y = a + b$ and $x^2 + y^2 = a^2 + b^2$. Prove that $x^n + y^n = a^n + b^n$ for all natural numbers n.

Solution. Prove that $x^n + y^n = a^n + b^n$.
We have $x^2 + y^2 = a^2 + b^2$.
Then

$$x^2 - a^2 + y^2 - b^2 = 0$$
$$(x-a)(x+a) + (y-b)(y+b) = 0.$$

Since $x + y = a + b$, it follows that $x - a = b - y$.
We obtain

$$(b-y)(x+a) + (y-b)(y+b) = 0$$
$$-(y-b)(x+a) + (y-b)(y+b) = 0$$
$$(y-b)[-(x+a) + y + b] = 0$$
$$(y-b)(-x - a + y + b) = 0.$$

Consequently, $\begin{bmatrix} y - b = 0 \\ -x - a + y + b = 0 \end{bmatrix}$ or $\begin{bmatrix} y = b \\ x - y = b - a \end{bmatrix}$.

In case, $y = b$, it follows that $x = a$. Then $x^n + y^n = a^n + b^n$.
In case, $x - y = b - a$, it follows that $\begin{cases} x - y = b - a \\ x + y = a + b \end{cases}$. Solve the equation system, we obtain $x = b$ and $y = a$.
Thus, $x^n + y^n = a^n + b^n$.

Problem 15. Given that $a > b > 0$ and satisfy $a^2 + b^2 = 2025ab$. Compute $A = \dfrac{a-b}{a+b}$.

Solution. We have

$$A^2 = \left(\dfrac{a-b}{a+b}\right)^2$$
$$= \dfrac{(a-b)^2}{(a+b)^2}$$

238

9.2. How To Solve It

$$= \frac{a^2 - 2ab + b^2}{a^2 + 2ab + b^2}$$
$$= \frac{a^2 + b^2 - 2ab}{a^2 + b^2 + 2ab}$$
$$= \frac{2025ab - 2ab}{2025ab + 2ab}$$
$$= \frac{2023ab}{2027ab}$$
$$= \frac{2023}{2027}.$$

It follows that $A = \pm\sqrt{\dfrac{2023}{2027}}$.
By knowing that $a > b > 0$, we obtain $A > 0$.
Consequently, $A = \sqrt{\dfrac{2023}{2027}}$.

Problem 16. Given x, y and z be three real numbers that satisfy $x + y + z = 0$ and $x^2 + y^2 + z^2 = a^2$. Find $x^4 + y^4 + z^4$ in terms of a.

Solution. We have $x + y + z = 0$. Then $y + z = -x$.
It implies that $(y + z)^2 = (-x)^2$ or $y^2 + 2yz + z^2 = x^2 = 0$.
It follows that

$$y^2 + z^2 - x^2 = -2yz$$
$$\left(y^2 + z^2 - x^2\right)^2 = (-2yz)^2$$
$$y^4 + z^4 + x^4 - 2x^2y^2 - 2x^2z^2 + 2y^2z^2 = 4y^2z^2$$
$$x^4 + y^4 + z^4 = 4y^2z^2 + 2x^2y^2 + 2x^2z^2 - 2y^2z^2$$
$$x^4 + y^4 + z^4 = 2y^2z^2 + 2x^2y^2 + 2x^2z^2$$
$$2\left(x^4 + y^4 + z^4\right) = x^4 + y^4 + z^4 + 2x^2y^2 + 2y^2z^2 + 2z^2x^2$$
$$2\left(x^4 + y^4 + z^4\right) = \left(x^2 + y^2 + z^2\right)^2.$$

Since $x^2 + y^2 + z^2 = a^2$, we obtain $2(x^4 + y^4 + z^4) = (a^2)^2 = a^4$.
Therefore, $x^4 + y^4 + z^4 = \dfrac{a^4}{2}$.

Problem 17. Compute

$$P = \left(1 - \frac{4}{1}\right)\left(1 - \frac{4}{9}\right)\left(1 - \frac{4}{25}\right)\cdots\left[1 - \frac{4}{(2n-1)^2}\right].$$

Solution. Observe that

$$1 - \frac{4}{(2k-1)^2} = \frac{(2k-1)^2 - 4}{(2k-1)^2}$$
$$= \frac{(2k-1)^2 - 2^2}{(2k-1)^2}$$
$$= \frac{(2k-1-2)(2k-1+2)}{(2k-1)^2}$$
$$= \frac{(2k-3)(2k+1)}{(2k-1)^2}$$
$$= \frac{2k-3}{2k-1} \times \frac{2k+1}{2k-1}.$$

It follows that

$$1 - \frac{4}{1} = \frac{-1}{1} \times \frac{3}{1};$$
$$1 - \frac{4}{9} = \frac{1}{3} \times \frac{5}{3};$$
$$1 - \frac{4}{25} = \frac{3}{5} \times \frac{7}{5};$$
$$\vdots$$
$$\text{and} \quad 1 - \frac{4}{(2n-1)^2} = \frac{2n-3}{2n-1} \times \frac{2n+1}{2n-1}.$$

Consequently,

$$\left(1 - \frac{4}{1}\right)\left(1 - \frac{4}{9}\right)\left(1 - \frac{4}{25}\right)\cdots\left[1 - \frac{1}{(2n-1)^2}\right]$$
$$= \left(-\frac{1}{1} \times \frac{1}{3} \times \frac{3}{5} \times \cdots \times \frac{2n-3}{2n-1}\right) \times \left(\frac{3}{1} \times \frac{5}{3} \times \frac{7}{5} \times \cdots \times \frac{2n+1}{2n-1}\right)$$
$$= \left(-\frac{1}{2n-1}\right)\left(\frac{2n+1}{1}\right) = -\frac{2n+1}{2n-1}.$$

Therefore, $P = -\dfrac{2n+1}{2n-1}$.

Problem 18. Suppose that x is a nonzero real number. Let $S_n = x^n + \dfrac{1}{x^n}$ for all nonnegative integers n. Prove that

$$S_{n+1} = S_n S_1 - S_{n-1}$$

9.2. How To Solve It

for all $n \geq 1$.

Solution. Prove that $S_{n+1} = S_n S_1 - S_{n-1}$ for all $n \geq 1$.
We have

$$\left(x^n + \frac{1}{x^n}\right)\left(x + \frac{1}{x}\right) = x^{n+1} + x^{n-1} + \frac{1}{x^{n-1}} + \frac{1}{x^{n+1}}$$

$$= \left(x^{n+1} + \frac{1}{x^{n+1}}\right) + \left(x^{n-1} + \frac{1}{x^{n-1}}\right).$$

It follows that
$$S_n S_1 = S_{n+1} + S_{n-1}.$$
Therefore, $S_{n+1} = S_n S_1 - S_{n-1}$ for all $n \geq 1$.

Problem 19. Let x be a positive real number such that

$$x^2 + \frac{1}{x^2} = 2.$$

Evaluate:

1. $x + \frac{1}{x}$;

2. $x^3 + \frac{1}{x^3}$;

3. $x^5 + \frac{1}{x^5}$;

4. $x^7 + \frac{1}{x^7}$.

Solution. Evaluate:

1. $x + \frac{1}{x}$
   We have

$$\left(x + \frac{1}{x}\right)^2 = x^2 + 2(x)\left(\frac{1}{x}\right) + \frac{1}{x^2}$$

$$= x^2 + 2 + \frac{1}{x^2}$$

$$= x^2 + \frac{1}{x^2} + 2.$$

From the hypothesis, $x^2 + \frac{1}{x^2} = 2$.
It follows that

$$\left(x + \frac{1}{x}\right)^2 = 2 + 2 = 4.$$

Since $x > 0$, then $x + \dfrac{1}{x} > 0$.

It implies that $x + \dfrac{1}{x} = 2$.

Therefore, $x + \dfrac{1}{x} = 2$.

2. $x^3 + \dfrac{1}{x^3}$

From Previous problem, we have
$$S_{n+1} = S_n S_1 - S_{n-1}$$
for all $n \geq 1$.

Taking $n = 2$, we obtain
$$\begin{aligned}S_3 &= S_2 S_1 - S_0 \\ &= 2(2) - \left(x^0 + \dfrac{1}{x^0}\right) \\ &= 4 - 2 \\ &= 2.\end{aligned}$$

Therefore, $x^3 + \dfrac{1}{x^3} = 2$.

3. $x^5 + \dfrac{1}{x^5}$

We have
$$\begin{aligned}S_4 &= S_3 S_1 - S_2 \\ &= 2(2) - 2 \\ &= 2.\end{aligned}$$

It follows that
$$\begin{aligned}S_5 &= S_4 S_1 - S_3 \\ &= 2(2) - 2 \\ &= 2.\end{aligned}$$

Therefore, $x^5 + \dfrac{1}{x^5} = 2$.

4. $x^7 + \dfrac{1}{x^7}$

We have
$$S_6 = S_5 S_1 - S_4$$

9.2. How To Solve It

$$= 2(2) - 2$$
$$= 2.$$

It follows that

$$S_7 = S_6 S_1 - S_5$$
$$= 2(2) - 2$$
$$= 2.$$

Therefore, $x^7 + \dfrac{1}{x^7} = 2$.

Problem 20. Given that a and b are two real numbers such that $ab = 1$. Prove that

$$a^{n+1} + b^{n+1} = (a^n + b^n)(a + b) - (a^{n-1} + b^{n-1})$$

for all positive integers n.

Solution. Prove that $a^{n+1} + b^{n+1} = (a^n + b^n)(a + b) - (a^{n-1} + b^{n-1})$. We have

$$(a^n + b^n)(a + b) = a^{n+1} + a^n b + ab^n + b^{n+1}$$
$$= a^{n+1} + ab(a^{n-1} + b^{n-1}) + b^{n+1}$$
$$= (a^{n+1} + b^{n+1}) + (a^{n-1} + b^{n-1}).$$

Therefore, $a^{n+1} + b^{n+1} = (a^n + b^n)(a + b) - (a^{n-1} + b^{n-1})$ for all positive integers n.

Problem 21. Given six nonzero real numbers a, b, c, x, y and z such that $a = by + cz, b = cz + ax$ and $c = ax + by$. Prove that

$$(x+1)(y+1)(z+1) = \dfrac{(a+b+c)^3}{8abc}.$$

Solution. Prove that $(x+1)(y+1)(z+1) = \dfrac{(a+b+c)^3}{8abc}$.
We have $a = by + cz, b = cz + ax$ and $c = ax + by$.
Adding all of the above equations, we obtain

$$a + b + c = 2ax + 2by + 2cz$$
$$= 2(ax + by + cz).$$

It follows that
$$ax + by + cz = \frac{a+b+c}{2}.$$
Since $a = by + cz$, it implies that
$$ax + a = \frac{a+b+c}{2}.$$
Then
$$a(x+1) = \frac{a+b+c}{2}.$$
We obtain
$$x + 1 = \frac{a+b+c}{2a}.$$
Similarly,
$$y + 1 = \frac{a+b+c}{2b}$$
and
$$z + 1 = \frac{a+b+c}{2c}.$$
Hence,
$$(x+1)(y+1)(z+1)$$
$$= \frac{a+b+c}{2a} \times \frac{a+b+c}{2b} \times \frac{a+b+c}{2c}$$
$$= \frac{(a+b+c)^3}{8abc}.$$

Problem 22. Given three nonzero real numbers a, b and c such that
$$\frac{1}{a} + \frac{1}{b} + \frac{1}{c} = \frac{1}{a+b+c}.$$
Prove that $a + b = 0, b + c = 0$ or $c + a = 0$.

Solution. Prove that $a + b = 0, b + c = 0$ or $c + a = 0$.
We have
$$\frac{1}{a} + \frac{1}{b} + \frac{1}{c} = \frac{1}{a+b+c}.$$
Then
$$\frac{1}{a} + \frac{1}{b} + \frac{1}{c} - \frac{1}{a+b+c} = 0.$$

9.2. How To Solve It

It follows that
$$\frac{a+b}{ab} + \frac{a+b+c-c}{c(a+b+c)} = 0.$$

We obtain
$$(a+b)\left[\frac{1}{ab} + \frac{1}{c(a+b+c)}\right] = 0$$
$$(a+b)\left(ac+bc+c^2+ab\right) = 0$$
$$(a+b)\left[a(b+c)+c(b+c)\right] = 0$$
$$(a+b)(b+c)(a+c) = 0.$$

Therefore, $a+b=0, b+c=0$ or $c+a=0$.

Problem 23. Given that a, b and c are three real numbers such that $a+b+c=0$. Prove that
$$a^4+b^4+c^4 = \frac{1}{2}\left(a^2+b^2+c^2\right)^2.$$

Solution. Prove that $a^4+b^4+c^4 = \frac{1}{2}\left(a^2+b^2+c^2\right)^2$.
Since $a+b+c=0$, we obtain $a+b=-c$.
Raise both sides of the above identity to square, it follows that
$$a^2+2ab+b^2 = c^2.$$

Then
$$a^2+b^2-c^2 = -2ab.$$

Again, raise both sides of the above identity to square, we obtain
$$a^4+b^4+c^4+2a^2b^2-2b^2c^2-2c^2a^2 = 4a^2b^2.$$

It implies that
$$a^4+b^4+c^4 = 4a^2b^2 - 2a^2b^2 + 2b^2c^2 + 2c^2a^2$$
$$= 2a^2b^2 + 2b^2c^2 + 2c^2a^2.$$

Adding $a^4+b^4+c^4$ to both sides of the last identity, we obtain
$$2\left(a^4+b^4+c^4\right) = \left(a^2+b^2+c^2\right)^2.$$

Therefore, $a^4+b^4+c^4 = \frac{1}{2}\left(a^2+b^2+c^2\right)^2$.

www.ingramcontent.com/pod-product-compliance
Lightning Source LLC
Chambersburg PA
CBHW031614210526
45464CB00004B/1580